HALLEY'S QUEST

A SELFLESS GENIUS AND HIS TROUBLED *PARAMORE*

Julie Wakefield

Joseph Henry Press

Washington, D.C.

Joseph Henry Press • 500 Fifth Street, NW • Washington, DC 20001

The Joseph Henry Press, an imprint of the National Academies Press, was created with the goal of making books on science, technology, and health more widely available to professionals and the public. Joseph Henry was one of the founders of the National Academy of Sciences and a leader in early American science.

Any opinions, findings, conclusions, or recommendations expressed in this volume are those of the author and do not necessarily reflect the views of the National Academy of Sciences or its affiliated institutions.

Library of Congress Cataloging-in-Publication Data

Wakefield, Julie.
  Halley's quest : a selfless genius and his troubled Paramore / Julie Wakefield.
    p. cm.
  Includes bibliographical references and index.
  ISBN 0-309-09594-8
  1. Scientific expeditions—Great Britain. 2. Halley, Edmond, 1656-1742. 3. Paramore (Ship) 4. Geodetic astronomy—Observations. 5. Geomagnetism—Observations. 6. Discoveries in science—Great Britain I. Title.
  Q115.W35 2005
  526'.6—dc22

                    2005016511

Dedication:

To my parents, Judy and Lou

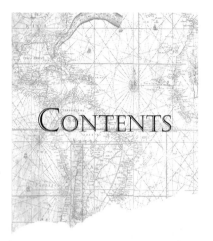

Contents

THIRD VOYAGE :
1701

FIRST VOYAGE:
1698–1699

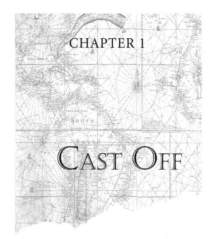

CHAPTER 1

CAST OFF

You are to make the best of your way Southward of the equator, and there to observe on the East Coast of South America, and the West Coast of Africa, the variations of the Compass, with all the accuracy you can as also the true Situation both in Longitude and Latitude of the Ports where you arrive.

Edmond Halley knew every detail of the royal decree. Though not as extensive as he had originally hoped—there was no round-the-world expedition, no probing into Pacific waters as he initially envisioned—it did authorize an unprecedented voyage. Halley would take command of a newly built 52-foot ship for the sake of science. He would touch the shores of at least four continents in a new kind of expedition where the real treasure would be clusters of numbers. His friend Isaac Newton was the man of gravity. Halley would be, at the very least, the man of magnetism. And in the intense emotion of the times, his ultimate success could chart a course in favor of a more modern future.

It had not been easy. Winning the unique commission, which also made him a civilian captain in the Royal Navy, ranked as a *coup de maitre* for Halley. Her Majesty, Queen Mary II, then in her early 30s but frail after several miscarriages, had approved the plan five

years earlier—in 1693—while staring down a mystery illness and
without the input of her husband, King William III. William had
again ventured abroad, dodging cannon fire as the war of attrition
against France waged on. The popular queen had accepted almost
all of Halley's recommendations in his royal petition. The orders
Halley finally received were almost as if the scientist had penned
them himself.

Halley had been contemplating such a voyage for years. His interest in magnetism and his earlier studies had convinced him that
obtaining a large-scale perspective of Earth's mysterious magnetic
forces would somehow answer the longitude problem that had been
plaguing sailors and wreaking havoc on the Royal Navy. Sovereignty
over the seas was the chief lure for the Crown. Halley had tempted
the powers behind the British monarchy to finance his ambitious
quest, which required turning half the world into his laboratory.

The whole globe at the time seemed to be in transition. It was
shrinking as knowledge was growing. Recent discoveries of science
had seemed to overtake those of new landmasses. But the English
Crown still had not lost its lust for potential geographic bounty. In
addition to sponsoring grand scientific inquiries, the royal decree to
Captain Halley explicitly mentioned the globe's final frontier:

> You are likewise to make the like observations at as many of the Islands
> in the Seas between the aforesaid Coasts as you can (without too much
> deviation) bring into your course: and if the Season of the Year permit,
> you are to stand so far into the South, till you discover the Coast of the
> Terra Incognita, supposed to lie between Magellan's Straits and the Cape
> of Good Hope, which coast you are carefully to lay down in its true
> position.

Terra incognita had inspired countless mariners. Soon after classical geographers first reckoned that Earth might be a giant sphere,
the possibility emerged of an unknown south land as vast as the continents in the Northern Hemisphere. Ancients like the famous Greek
astronomer and geographer Ptolemy reasoned that the planet would

certainly tilt without its balancing effect. Such an expanse was sure to harbor riches beyond imagination.

The passing of centuries only burnished the myth. Venetian vagabond Marco Polo professed to have witnessed *terra incognita*'s gold-filled hills and bountiful wildlife from afar in the 13th century. In May 1606 the Spanish captain Pedro Fernandez de Queiros claimed to have set foot on the continent. In the ensuing decades, Dutch East India men purported to have periodically made landfall on its western coast on jaunts past the cape on the journey to the Spice Islands. Although such claims attracted skepticism, the world's interest had been piqued.

By Halley's day, mapping *terra incognita* had become equated with finding the Holy Grail. Countless men had died in the pursuit. *Terra incognita* had grown from a mere landmass to a symbol of the future's magnificence. A place with untouched waters and treasures, never-tasted flavors and elixirs, and colors more vibrant than real life—a mystical Eden where open skies and endless possibilities still reigned. Near the 18th century, even the practical-minded Halley was not immune to its allure. *Incognita*'s inclusion in the English domain played to the queen's fancies.

Halley had been more than willing to comply. At heart he was a great adventurer, passionate about all realms of discovery. But he was savvy enough to know that more than mere scientific merits were required to secure the queen's approval. He had to win her personal interest. He also knew that fulfilling her ambitions could catapult him into even more powerful realms of London society, garnering him not only intellectual credentials unobtainable by sheer toil at the universities of Oxford or Cambridge but also social status beyond that of his birth right as the son of a wealthy merchant and property owner.

Halley, properly discreet to Her Majesty, had made little mention of the clever politicking behind the world's first scientific mission. Almost every detail of his preparations, including the christening of his ship, the *Paramore*, a 52-foot pink, was working out as the methodical and analytical thinker had diligently planned.

Whether Halley met or corresponded personally with the queen, as he did with her father, King James II, several years earlier, is unrecorded. The scant diaries and letters of Queen Mary that survive do not mention Halley or the *Paramore*. But as coleader of an island empire, she showed an abiding respect for mariners. In a journal of 1692, she had believed that "the fate of England . . . depend[s] on our success at sea." She added: "If our fleet was beaten, we knew there was an army ready to devour us from abroad." When a petition for the unusual voyage, endorsed by Halley's learned friends in the Royal Society of London, was laid before Mary in 1693, it was reported that "the Queen Her Majesty is graciously pleased to encourage the said undertaking." She commanded, through the Admiralty, that a ship of about 80 tons be built in "their Majesty's Yard at Deptford as soon as may be." Deptford, on the south bank of the Thames west of Greenwich, was the nation's main naval dockyard where food, rum, clothing, and other provisions were warehoused.

In a complementary task, the monarchy also directed Halley to the Caribbean Islands on his way home to determine the location of England's tropical plantations. But the inclusion in the decree of his plan to survey unseen magnetism across the oceans made Halley most proud. His instructions specifically stated: "In all the Course of your Voyage, you must be careful to omit no opportunity of Noting the variation of the Compass, of which you are to keep a Record in your Journal." The words gave official license to an indefatigable passion that had been burning inside Halley. His ideas on how Earth's magnetism might solve navigational problems had so impressed the Royal Society that its patrons wasted no time in advancing his proposal. The monarchy had never backed a venture like this before, and Halley was fully conscious of the significance of the moment.

Halley had long been taken with magnetism, a phenomenon as equally puzzling and seemingly as universal as gravity. As the 18th century was dawning, this fickle force of attraction and repulsion still mystified the world's finest minds. For a start, why did the quivering magnetic compass stray from aligning with the North Star at various locales on the globe? The mystery contributed to the challenges of navigating at sea, which were many, and hindered the island kingdom's prosperity. Whether by choice or not, one out of every hundred Londoners was a mariner. Solving this empire-enhancing conundrum comprised the largest part of Edmond Halley's charge.

But before the journey could even begin, a smallpox epidemic, showing no deference to royal blood, claimed Halley's benefactor in December of 1694. The devout queen had lived 32 years, long enough to see construction of the expedition's vessel, the *Paramore*, completed and launched. Now Mary lay unmoving amid ermine and purple robes. It is likely that not even the sublime music that Henry Purcell wrote for the mourning of the great queen could soothe Halley's doubts.

Although the scientist had already received his commission for the vessel, the new ship was shelved and the mission put on hold. (Of course, the term "scientist" wouldn't be coined for more than a century at a meeting of the British Association for the Advancement of Science in 1834. In the 17th century, such men were known as "natural philosophers.") Just three years before, Halley had been denied by religious authorities the prestigious Savilian Professorship of Astronomy at Oxford. Now, in an even greater setback, his most ambitious quest seemed as dead as its royal matron.

The queen's Dutch husband, King William, had been indifferent to London, its politics, and frequently the queen herself. He'd once even said, "Let the Queen rule you, she is English, she understands what you want, I don't." A continental man who still spent long summers in Holland, the aloof—albeit-underrated—king was derisively nicknamed "Hooknose" by the English.

But William could not have shown more devotion to his wife's memory. Long past the customary mourning period, he continued to wear, above his left elbow, a black ribbon tied to a golden ring that contained some of Mary's rich brown hair. More importantly, he continued the queen's grandest plans. Besides converting the royal residence at Greenwich into a naval hospital for injured seamen, he made sure English America's second college arose off the tidewater flats of Virginia and that it was designed—as the queen had desired—by no less an intellect than Sir Christopher Wren, another Halley confidant. In addition to presiding over the resurrection of London after the Great Fire, Wren had redesigned the Earl of Nottingham's mansion, which would become known as Kensington Palace, for the royal couple soon after they came to power. (Their initial residence at Hampton Court Palace proved too far from London.) The new Virginia college, William and Mary, forever paired the king and queen, and was founded as Mary had intended—to train missionaries in the New World.

Yet questions soon arose as to whether King William would continue to pursue the queen's main scientific interest. Or would this foreign ruler fall prey to traditionalists in British society? With elegant wit and caustic slogans, more conservative intellectuals were denigrating scientific reasoning and belittling the nation's inclination into what they saw as a godless rush into the future.

Newton's mathematical logic and similar radical papers by Halley and other Royal Society thinkers had provoked an academic backlash and a controversial "Battle of the Books," over the extent to which modern discoveries had superseded classical learning. Sir William Temple, intimate of monarchs, led the opposition, along with the literary likes of Jonathan Swift. Swift called the Royal Society intellects "the Atheists of the Age." This public outcry only added to Halley's doubts about whether his idle *Paramore* would ever sail for the sake of science. It is easy to imagine Halley in his late 30s, wide-eyed, tossing at night over the thought of such stalwarts snipping his ambition for intellectual glory.

But Halley never abandoned all hope. Even some 16 months after the queen's death, he felt possessive about the neglected ship. One Saturday in the spring of 1696, he postponed a call on Newton's home, apologizing because he felt "obliged to go on board my frigate," even though the *Paramore* was unsailable in wet dock. The next day, with prospects still bleak for the science voyage, Halley talked to Newton about a job. Newton had recently arrived in London from Cambridge to become master of the mint. Would Halley like to be a deputy in Chester? Forced to face the prospects that his mission really might never happen, Halley accepted, even though the lucrative job, specializing in silver coinage, would put him more than 200 miles away from London's vibrant mix of a half million souls.

Finally in 1698, nearly four years after Queen Mary's death, King William revived the *Paramore* mission. It was probably the signing of the Peace of Ryswick in 1697 ending the war with France, then Europe's most powerful country, that caused William to decide it was safe to allow the *Paramore* to set sail, as the English Channel and the Atlantic Ocean would no longer be infested with enemy warships. By early August the *Paramore* was being planked from the waterline down in a protective sheath to guard against the tunnelings of the teredo worm. All signs pointed to a long mission in blue waters.

IT WAS OCTOBER 15 WHEN THE monarchy at last tasked the unusual science voyage. For the first time, Captain Edmond Halley greeted his lieutenant and the 18 crew members who were to share the ship's cramped quarters for untold months. It is a shame no one recorded the expressions on their faces. To Halley all but two of the men were complete strangers. Did these veteran sailors stifle their surprise at being led by a pale-skinned gentleman who looked more suited to parlor debates? The crew that would leave Deptford counted John Dunbar as midshipman, John Dodson as boatswain and gunner, and Thomas Price as carpenter. Halley found no objections to them, nor to the gunner's mate Mathew Butts; the carpenter's mate, William Dowty; or his own clerk, Caleb Harmon. The seven seamen—Peter Ingoldsby, John Thompson, James Glenn, David Wishard, Samuel Withers, Thomas Davis, and John Vinicot—were undoubtedly sound. And Halley didn't gripe about the choice of the ship's servants: Richard Pinfold for Halley, John Hodges for the boatswain, Robert Dampster for the carpenter, and Thomas Burton.

These crewmen were not the usual petty criminals or souses rounded up from local pubs, which was the recruitment method often favored by privateering outfits to supplement their numbers. Halley had given some careful thought to the Navy Board's selection of his crew. Even for the most seasoned commanders, discipline could be challenging on a long voyage. To offset his dearth of experience and guard against mutiny, Halley had filed two requests: first, that the entire crew be paid from the king's coffers, and next, that his second in command be a commissioned officer in the Royal Navy. On that last count, he would regret not being more specific.

The only shipmate Halley had personally requested by name was naval surgeon George Alfrey. Although small ships typically didn't require or merit the distinction, Halley knew that the health of his lean crew was vital for this mission, expected to last 12 months.

As if the tension of being a gentleman captain were not enough, the second in command selected by the Navy was familiar by name to Halley: First Lieutenant Edward Harrison. Halley's heart must have dropped. Edward Harrison was notorious to Halley for his scholarly pretense! With a sense of either pride or irony, someone probably in the Admiralty had selected the veteran sailor and one-time author to accompany Halley. It could have been totally innocent. Lieutenant Harrison had developed his own scientific ideas about the pressing problem of calculating longitude at sea. His ideas not only clashed with Halley's but also set him at personal odds with his captain. Only two years before, Harrison had published a thin book in London titled *Idea Longitudinis: Being, a Brief Definition of the Best Known Axioms for Finding the Longitude.* In its pages he posited ways to solve the troublesome location problem. He wrote that he, as a sailor, "may prove to be a more Competent Artist in Navigation" than any gentleman mathematician. So highly had Harrison regarded his own findings, he had submitted them to the Admiralty, the Navy Board, and the Royal Society. But the reviews had come back uniformly poor. Standing tall among the critics had been none other than Edmond Halley.

Whether the Admiralty had appointed Harrison believing that one of the Navy's own native intellects would be best suited for a gentleman captain from the Royal Society or whether it was a wry jest—akin to putting two cocks in the same cage—will never be known. Maybe it was a little of both.

For his part, Halley realized he could never compete with Harrison from stem to stern. He had even openly acknowledged his lack of nautical experience: "Perhaps I have not the whole Sea Dictionary so perfect as he." But his admission had only fueled further resentment.

In the Restoration Navy of England, the world's most powerful, two classes of senior officers were deeply entrenched. Experienced sailors, known as tarpaulins, had struggled their way up to command. Then there were those who had risen quickly despite their shaky sea legs: gentleman commanders. Historically, the two clashed like discordant waves. Gentleman seamen thought the tarpaulins rough and uncouth. The salty tarpaulins, in turn, scoffed at the lack of nautical know-how of such well-connected aristocrats. In his 1617 *Discourse on Pirates*, for example, Sir Henry Mainwaring contended that handling a ship with "discretion and judgment, to manage, handle, content and command the company, both in fear and love (without which no Commander is absolute)" exceeded a gentleman captain's capacity.

Although sea captains had the absolute authority of a dictator at sea and Halley was renowned as a rising scientific star, his veteran crew seemed predisposed to dismiss the civilian Halley. His pallid and still mostly smooth skin, a far cry from the creased leather of most mariners, reeked of prissy amateurism. The polite manner that served him so well in society could prove a liability in rough seas.

THE UNPROVEN VESSEL moved out from its mooring among the dense groves of masts on the Thames around noon on October 20, 1698. This bulge downriver from the bridge, known as the Pool of London, boasted some 2,000 ships at the time of Halley's departure, according to Daniel Defoe, who would soon write that tale of a marooned mariner, *Robinson Crusoe*. Halley had stowed the royal decree securely in his cabin, just in case proof of his mission and authority would ever be needed.

Ahead loomed any variety of unknowns: raging seas, barbaric pirates, fickle winds. Behind trailed the still-perceptible wake forged by years of frustrating preparations and politics. At home he would leave a pregnant and dear wife of six years, Mary Tooke, and two daughters. If Halley harbored any reservations about leaving his young family for such a sojourn, he didn't let them show. If nothing else, his ambition had curbed such concerns.

Several days passed before the *Paramore* had cleared the Thames and entered the English Channel. The channel provided the outbound crew with time to learn the ropes together before hitting wide-open water. Captain Halley had been tasked to improve navigation over much of the globe. The monarchy trusted him to venture south of the equator to observe the variations of the compass off the east coast of South America and the west coast of Africa and to find the latitude and longitude at each port he visited. But his anxious crew had a more pressing concern: Could this man more at ease in the society of powdered periwigs navigate this single 52-foot ship?

Whether his crew was aware of it or not, Halley at the turn of the century numbered among the world's greatest navigational talents. When it came to deciphering a pathway from the positions of the stars and planets, he was first class. Some prominent contemporaries readily acknowledged that his scientific prowess exceeded that of any seaman at the time. London diarist Samuel Pepys, a Royal Society president, complimented Halley on his rare blend of gifts as the first Englishman—and possibly man of any origin—to be competent in both the "science and practice of navigation." Even among the elite cadre of scientists of the time, practitioners of the so-called new philosophy, it was rare to find one who so mixed theory and experiment, civic-mindedness and pragmatism, and tied it to the real world, as the relatively selfless Halley did.

Long after his death, Halley would be popularly remembered for a far lesser feat, his astronomical acumen in applying Newton's gravitational theory to predict the return of a lustrous comet. But at this time the 42-year-old Halley would need more than otherworldly smarts to complete his most challenging mission. The world's first official voyage for science, this possible usher of society's new faith in human ingenuity, on its face seemed a recipe for disaster.

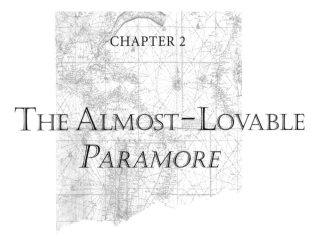

CHAPTER 2

THE ALMOST–LOVABLE
PARAMORE

66 "All the Ships out turned us and went away before us," Halley logged on October 29, 1698. What should have been a triumph—the full flutter of sails, the strained rigging, the smell of adventuresome brine replacing the staleness of domesticity—subsided into immediate disappointment. Halley's boat was barely seaworthy!

The early trials and tribulations with the *Paramore*, many already evident even before leaving the English Channel, betrayed the shortcomings of these types of wooden vessels. They leaked, they struggled to sail close to the wind, and they were unsteady. Maintaining proper ballast alone challenged even the most sage captain. Gusts easily fractured masts and yards. Some seagoers affectionately called them pinks, but probably not Halley, at least not at this time. He had not established any faith in the vessel in the early goings.

Originally of Dutch design and named from the Middle Dutch "pincke," these ships as a class were slow, plodding sailers. Their square rigging only compounded their sluggishness. On the plus side, the three-masted vessels had a large cargo capacity for their size, ideal for safely stowing Halley's bulky and fragile cache of scientific instruments. Compared with sleeker sailers from the same class, pinks offered considerable maneuverability in shallow waters close to shore. Along with the 50 Men-of-War used by His Majesty William of Orange, the Protestant champion, to cross the channel and to conquer England in 1688 were some 500 transports, including 60 pinks. No doubt pinks were better suited for traversing short distances in a protected channel than voyaging thousands of miles on the Atlantic chop.

Among this class of lemons, the *Paramore* had proven especially sour. Her shallow draught—roughly seven feet deep—offered poor stability, allowing ready movement from side to side. She was not only shallower but a good 20 feet shorter from stem to stern than the typical merchant ship of the day. Meanwhile, her broad beam—spanning 18 feet from port to starboard—hindered her forward progress, especially as barnacles and algae would amass on the hull. Halley had scrawled in his log, probably gritting his teeth, that she went more to leeward than most other ships. So awkward was her movement that Halley thought she looked the part of an ungainly renegade ship. His observations were not unwarranted. The appearance of the poor ship would later jeopardize her crew.

Shortcomings in Halley's particular boat were all the more ironic, for the vessel had been custom built explicitly for the mission. And it had been crafted not by apprentice builders but by a master shipwright at Deptford named Fisher Harding. She was the only ship in the royal fleet ever to bear the name *Paramore* and one of only two pinks ever held by the Admiralty. Displacing 89 tons, she had launched April 1, 1694, in time for what would be Queen Mary's last birthday. And then the long wait had begun.

Halley had successfully garnered his dream commission but not by any means his dream ship. Captaining a pink was far from his first choice. At various points in his preparations, Halley had contemplated securing a better class of ship. But his informal inquiries had been brushed aside.

BY MID-NOVEMBER, 1698, the *Paramore* had only made Portsmouth in the English Channel—about half the distance from London and the open ocean off Land's End. True to form, the *Paramore*'s defects had already impeded the mission. The sand used for ballast had clogged the hand pumps, which were used to drain seawater that seeped inside the hull. Halley had wisely decided to head into port to replace the gritty sand with shingle ballast. Such small, water-worn stones would more effectively stabilize the ship while enabling water to pass through them without plugging up the pumps.

Though undoubtedly distressing to Halley to have to call to port so soon after launch, the *Paramore*'s unplanned stopover in Portsmouth harbor proved a welcome twist of fate. While workers replaced the unacceptable ballast with shingle, the gentleman commander turned his attention to the broader vision of his mission and began the painstaking process of gleaning information from the skies and Earth's magnetic field. Independent of each other, Halley and Harrison jotted down observations of conditions and the compass and other notes in their journals. The mission's first measurements of magnetic variation had come at Portsmouth. The phenomenon, also referred to as magnetic declination, is the difference between true north and magnetic north. Halley defined it as "the deflection of the Magnetic Needle from the true Meridian."

Measuring magnetic variation was first put forth as a way to find longitude shortly after the Italian Christopher Columbus sailed the Atlantic in 1492 for King Ferdinand II and Queen Isabella of Spain. Scholars are unsure who first proposed the idea. Halley believed the observable degrees that a compass needle diverges from the astronomical north-south line could be linked to longitude—and without the need for knowing what time it was in two places at once: on ship and at the home port, which competing methods required. During Columbus's voyage, the magnetic declination had worked to Spain's advantage, causing his ships to sail farther south than intended and sight land in the knick of time—before the deadline by which he promised his crew they'd turn back if unsuccessful—days sooner than if they'd sailed toward geographic west. Inadvertently, Columbus had crossed the line where the variation of the magnetic pole is zero or where the magnetic and geographic poles coincide. After that meridian, when he thought he was heading due west by the compass, he was actually trending south of west.

More important to immediate concerns, though, Halley made the acquaintance of Rear Admiral John Benbow at the port. Although the main threats now in the channel were winds and tidal currents, the open seas were essentially lawless wilderness. The *Paramore* didn't have much to repel attacks. Stowage for scientific instruments outweighed that of weapons and ammunition. The ship had only six three-pound guns and two smaller guns on swivels. Despite all his planning, Halley had not resolved how to pass unscathed through pirate territory off North Africa's coast. "Our people were somewhat doubtful of going alone, for fear of meeting with a Sallyman," Halley wrote at the time. (Sallee was then a thriving port city on the Barbary Coast, now part of Morocco.)

Admiral Benbow, stuck in port awaiting favorable winds, had been recently appointed commander in chief of the king's ships in the West Indies. Impressed with Halley's quest, he agreed to escort the small *Paramore* past the African coast with his flotilla of warships bristling with cannons.

Finally the winds changed. The *Paramore* rode a moderate gale out of Portsmouth harbor on November 22 and soon joined Benbow's fleet off the Isle of Wight. The *Paramore* fired a five-round salute. The admiral flattered Captain Halley with a return salute of as many volleys, ignoring military protocol for a subordinate.

Ten big bangs and the best escort anyone could hope for. "If we can keep the Admiral Company those [pirate] apprehensions are over. He has promised to take care of us," Halley wrote to the Admiralty. Perhaps luck was finally turning in the *Paramore*'s favor.

CHAPTER 3

BATTLE OF THE BOOKS

W hen Halley's *Paramore* set sail, turn-of-the-century London was prospering amid an incredible bloom of ambition and intellect. In fashionably seedy parts of the city, coffeehouses buzzed with contentious questions of the day. Famous wits and poets like John Dryden mingled with scientific elite of the Royal Society, politicos, dilettantes, and others searching for inspiration or just the latest news and gossip. They frequented such haunts as the Prince of Orange, Muss's, and Garraway's Coffeehouse near the Royal Exchange. Most popular with Royal Society regulars like Halley was Jonathan's, according to the diary of leading experimenter Robert Hooke, off Cornhill and Lombard streets. Imbibing Turkish coffees and blowing whiffs off long clay pipes, the regulars talked about more than just the affairs of the Crown or the colonies. Arguably the greatest book since the Bible had recently been published by Halley's temperamental friend, Isaac Newton. It audaciously embraced much of the scope of Genesis. In his *Philosophia Naturalis Principia Mathematica* (or *Principia* for short), Newton claimed to disclose the underlying order of the universe—from the fall of an apple to the surges of tides, from the orbit of Jupiter to the blazing ominous trails of comets.

"No longer does error oppress doubtful mankind," said Halley, who had shepherded the work. "Things which so often tormented the minds of ancient Sages, we perceive." It was almost like he sounded a trumpet to herald a new age.

London had recovered from its Great Fire of 1666, but several planned ornaments, such as Sir Christopher Wren's domed St. Paul's Cathedral, had yet to dominate the skyline. King William III of Orange and Queen Mary II comfortably shared the power of the Crown. Their unusual double coronation coincided with the publication of John Locke's two treatises on government. That 1689 changeover in power altered thinking about individual liberties and politics and soon came to be called the Glorious Revolution and produced a bill of rights and law common to all. Despite all this—or maybe because of it—London society roiled with contradictions.

Social leaders espoused the values of virtue, public spirit, and liberty, but public corruption and private vice were rampant. As political pamphleteer Edward Stephens described it in a pamphlet circulated that year, William had freed England from the "abomination of popery," but "debauchery and impiety remained." William and Mary took his commentary seriously enough to officially respond to it in writing in 1690. They handed down proclamations against vice, immorality, and corruption. Societies soon formed to stifle bawdy houses and the like in the Tower Hamlets, a borough near the Tower of London. The Society for the Reformation of Manners, for one, paid informers to rein in private vice. Such societies themselves were conflicted: Their members valued prosperity and pleasure for themselves while relegating other humans idle and useless.

Soon after the *Paramore*'s send-off, postings from the East Indies in late November relayed that a London privateer known as Captain Kidd, once supported by King William, had gone pirate and captured a rich Arabian ship and Portuguese and Dutch ones as well. "But 'tis hope this advice is not proved true in all circumstances," one article concluded.

The most eagerly followed news announced the arrivals and departures of ships in London. Published no more than three times a week and mainly circulated in coffeehouses, newspapers also carried accounts of mutiny and piracy at sea with a frequency that made such treachery seem commonplace. People were not afraid to show their displeasure—and newspapers wrote about it.

Sir Francis Child was sworn in as lord mayor of the City of London in early November 1698 before the Barons of the Exchequer. Child went by water to Westminster (then outside London's walls) and returned to the Guildhall, the seat of government of the City of London, with the usual ceremonies. The civic rite didn't go as seamlessly as William and Mary's coronation. "There was no squibs thrown but the mob was very rude and threw dirt upon the balconies," according to the *Post Man*, a London two-page tabloid.

Details from the lives of Europe's sundry royalty also garnered top billing in the London rags, which besides the *Post Man* included the *London Gazette*, *Post Boy*, and the *Flying Post*. Reports from Paris, Lisbon, and Europe's other great cities were prominently featured. The papers also gave considerable ink to newly published books, revealing London society's respect for literature of any caliber. Given that most issues ran merely two pages, even a brief mention was significant. The first daily paper, the *Daily Courant*, wouldn't be launched for another four years.

Advertisements that trailed the news accounts provide a random cross section of London at the time Halley set sail. A promotion for a looking glass to detect kidnappers hyped the said device as "an infallible method to prevent the notorious villains in the future." It could reveal "their hellish and odious intrigues, monstrous designs, and notorious villainies in their base, though cunning wheedles and cursed inventions to entrap, decoy, and delude men, women, and children to fell them and be made slaves or into strange countries for base gain and lucre fate." There was a pitch for insurance for widows from the Office of the Society of Assurance for Widows. Other ads touted such health remedies as *Tinctura Caelestis*, "the celebrated medicine of the age," for gout, apoplexy, and rheumatism, and as Capital Salt, "an admiral remedy for diseases of the head such as vertigo or migraine, headache, hypochondriac passions, and vapors."

The Royal Society also shamelessly plugged offerings from its flagship journal. For example, around the time of Halley's departure on the Thames, the society ran an ad touting its exploits in an October issue of the *Post Man*. It began: *"Philosophical Transactions* giving some account of the present undertakings, studies and labors of the ingenuous in many parts of the world, continued by Dr. Hans Sloane, Secretary to the Royal Society . . ." and touted letters concerning Roman antiquities found in Yorkshire and another on the characteristics of some Indian plants. (Sloane was already a noted physician and naturalist.)

WITH ITS INTERNATIONAL DISPATCHES and sea trade, its mix with the Dutch, French, Spanish, Americans, and Germans, the West and East Indies, London was a flagship city of the world. Halley was born and bred in that metropolis of opportunity, a city where ties with London global trading companies could make a career. For Halley, his connections with the Levant Company in particular, which helped him secure voyage in 1676 to St. Helena, were forged through his marriage and other acquaintances about Winchester Street, a neighborhood an easy walk from the Tower of London.

Halley spent his boyhood in this core ward of the city where London's most successful merchants lived. His father was a soap boiler who set up shop and kept at least one family home in this prime neighborhood and owned a substantial number of other city properties. He was a well-to-do freeman of the Salter's Company, which was one of the most politically powerful London companies at the time.

Halley was a very sociable personality, scholars believe, forming associations with prominent Londoners from many different reli-

gious and political persuasions, including nonbelievers and the pi-
ous, Whigs and Tories alike. Members of the Levant Company, on the
one hand, were mainly Whigs, affiliated with the more liberal political party that strongly supported the installment of William and Mary
as rulers. Many of his other associates were reputed Tories or generally conservatives, some of whom also supported the removal of
James II in the so-called Bloodless Revolution in 1689. In the con-
fused December days in 1688 before the convention voted to grant
Prince William of Orange the English throne, Halley had socialized
with reputed Tories Robert Hooke and Christopher Wren at their favorite coffeehouses. His eclectic mix of patronage and connections
likely helped him secure means for his major scientific endeavors,
including his journey aboard the *Paramore*.

And it was there in London where he first glimpsed the ominous
comet of 1680 that would inspire both Halley and Newton to achieve
greatness. Many viewed this bright comet, visible to the naked eye, as
an unmistakable warning flare from God. Divine portent or not, it
was awesome by all accounts. The fiery comet seemed to appear from
nowhere and then fade just as quickly over a period of weeks. Not
only its path but also its very physical nature was a mystery. Halley
witnessed the quirky orbiting object in November just before he went
on a tour of France and Italy (a rite of passage for young gentlemen as
part of their education). He traveled with another Royal Society fellow named Robert Nelson, a colleague from his childhood neighborhood who became a religious writer and philanthropist and friend
for life.

En route to Paris, the duo saw the mysterious comet again. On
reaching the city of lights, Halley immediately tracked down the
French-born, Italian-educated Giovanni Domenico Cassini. The re-
vered astronomer was made director of the Paris Observatory in 1669
and would remain so for life. After reading an elaborate letter of introduction from England's Astronomer Royal John Flamsteed,
Cassini welcomed Halley, and they recorded a series of observations
on the comet tracking its path as best they could through the stars. To

Halley, Cassini's instruments and facility, which was completed in 1672 at the behest of King Louis XIV, paled in comparison to those at the disposal of Flamsteed at the Greenwich Observatory, finished three years later in 1675. (Sited on a picturesque knoll then several miles outside London and founded "for perfecting the art of navigation," the stargazing station was England's first official science building. The incarnation of the rival observatories ushered in a new era in astronomy. Before these observatories were built, observers had to build, equip, and fund their own facilities. Both institutions would prove instrumental to Halley's calling.)

Despite his partiality to Greenwich, Halley valued his stay with Cassini, meeting French savants like Henri Justel, then secretary to the Sun King, Louis the XIV. Halley noted in a letter to Robert Hooke that at the time "the general talk of the virtuosi here is about the Comet." Justel would soon emigrate to England, seeking religious freedom. There he was tapped by William's predecessor to be Charles II's librarian.

At the time of Halley's tour, Louis XIV was in the process of moving his court wholesale to Versailles. Abandoning the centrally situated Louvre palace by the Seine, he sought to be as separate as possible from the Parisian masses. At Versailles he would impose a rigid court etiquette that virtually reduced the nobility to mere courtiers. To further protect his power base, he elevated more easily manipulated commoners to posts as ministers and regional governors. Amid the extreme opulence of the gilded palace and its expansive gardens, more than 500 cooks prepared his meals and 4,000 servants catered to the French king's every whim. Meanwhile, his dragoons were busy harassing French Protestants.

Halley took up the mystery of the 1680 comet again just before his *Paramore* voyage. This time around, some 15 years after his first observation, he could try to apply some of Newton's theories to the problem. But when the *Paramore* was finally ready to sail, Halley put all else aside. The completion of his work on the orbits of comets would have to wait.

IF THE CUSP OF THE 18TH CENTURY entailed an ascendancy of science, a time when natural history began to fully challenge sacred history, then contemporary thought finally was rivaling, if not surpassing, the best of ancient Greeks and Romans. And Halley was determined to enlarge his role in this transformation. His drive to take science from the mere theoretical realm was intense enough for him to endanger his life at sea, a risk completely unnecessary for the financial success of a man of his class and stature. Yet the nonmonetary rewards seemed worthwhile. Halley sensed that many of the ideas that were around during this time would build the framework for modern thought. The roots of such future disciplines as archaeology, natural history, geology, and paleontology and workable systems of classification and taxonomy were emerging to guide the way knowledge is organized and used for centuries to come.

Not surprisingly, in this transitional time, which would be defined almost a century later as the early Enlightenment, London's public opinion was volatile and very much up for grabs. The papers of Royal Society thinkers were often so radical they prompted at once popular interest and criticism. In this tumultuous social clime, a series of disputes over the very essence of knowledge erupted. Chief among them, the so-called Battle of the Books busied London's printing presses. Simply put, the battle among English gentlemen pitted old against new, ancients versus moderns. The struggle would determine whether the ways of the past or those of the present should exert more influence on the future.

On the side of modernity were the Royal Society intellects of Halley, Newton, Wren, Hooke, and more. Such titans of science were beginning to establish a foothold in London culture. A group of 12 men had officially founded the Royal Society, London's eminent scientific group, in November 1660 after a lecture by polymath Wren at Gresham College. Wren, a mathematician, astronomer, and architect, is often credited with designing, among other things, the Royal Observatory at Greenwich, but it was actually the work of Hooke. Wren, however, picked its hilltop site. Twenty years earlier at Oxford, Wren had forged advances in the telescope, such as improvements to micrometers invented by William Gascoigne, that transformed the telescope's use from a tool for making qualitative observations to one capable of astronomical measurements. The movable wire scales in the eyepiece, for example, enabled a viewer to measure angles necessary to discern planetary positions from nearby stars. The Royal Society met weekly to witness scientific experiments and to discuss the new philosophy and assorted science topics. Charles II issued the first charter for the society in 1662. Unlike the Academie des Sciences de Paris, established in 1666 by King Louis XIV chiefly to improve maps and charts, Royal Society members neither received salaries from the king nor were they directed to solve specific problems. They were free to pursue individual interests and passions. For a time after the London fire of 1666, the society met at Arundel House, the London home of the Duke of Norfolk.

Since its beginnings, criteria for being nominated for election to the all-London club were rather vague. Reinforcing a system of patronage of the sciences for the first two centuries of its existence, wealthy amateurs and important men from church and state would fill the ranks along with meritorious scientists. (Members of the aristocracy were essentially automatically accepted.) Francis Bacon and his experimental natural philosophy, the "new philosophy" or system based on empirical and inductive principles and the active development of new arts and inventions, greatly influenced the original fellows—dilettantes and savants alike. Experimental philosophers strove to use science to produce practical knowledge as a means for bettering humankind, goals to which Halley aspired.

Early in his career in 1678, the Royal Society honored Halley as a fellow. He joined the likes of founders Robert Boyle, his one-time sponsor; Robert Hooke, the society's first curator of experiments; and Locke, a scientist in his own right. Boyle was known for experiments in which he used an air pump to unravel the laws of mechanics, heat, and air pressure. He sought evidence of God's will in the intricate workings of nature, which he dubbed "argument by design." Besides presenting experiments at society meetings, Hooke made significant contributions to microscopic observation, among other things. In 1665 he published his key work, *Micrographia,* which laid out the philosophy of "mechanism," whereby the world was assembled like a pocketwatch by none other than God, "the Great Mechanic." To Hooke the quest for natural knowledge entailed the discovery of new instruments to precisely probe naturally occurring phenomena. Many also knew Hooke for his discovery that the cork plant has a honeycomb structure of little chambers, which he dubbed cells.

The opposing faction of literati raging against the modernity machine included the formidable pen of Jonathan Swift. The clan of wits would later be joined by Alexander Pope, the mostly home-schooled London poet and satirist who came to epitomize English neoclassicism. Raised in Dublin, Swift first came to London in 1688, when the anti-Catholic revolt roiled Ireland, to live with Sir William Temple, a relative of his mother. Swift, somewhat begrudgingly, would serve on and off for the next decade, between jaunts back to Ireland, as secretary to the diplomat who had negotiated the marriage of then Prince William of Orange with the English Princess Mary.

In England the battle of intellects began in 1690—two years after a similar controversy ignited in France. Temple, then in his 60s and an established confidant to several English kings, published *Of Ancient and Modern Learning,* an essay that promoted classical methods over modern. It touted older works like *Aesop's Fables* and the *Epistles of Phalaris* as among the best literature available and disparaged modern arts and sciences.

Temple contended that "printing has increased copies rather than the quality or number of great books; that the earliest philosophers were the best; and despite the claims of Descartes and Hobbes no modern ones have excelled [surpassed] Plato and Aristotle." Temple's criticisms of modern thought extended to these vain scientists as "busying a man's brains to no purpose." In fact, modern "science had produced nothing to vie with the ancients, unless it be Copernicus's Theory [that the Earth revolves around the Sun, and not the other way round], or [the anatomist William] Harvey's [discovery of the] circulation of the blood. But since these hypotheses have not changed [the broader practices] of Astronomy or Medicine, they are of little use. Moreover, the earlier age was superior in music, rhyming, magic, and architecture. In fact the only great discovery of modern times is the loadstone, whence has come [modern] navigation and [greater] geographical knowledge [such as the discovery of America, a continent wholly unknown to the Ancients]. But even here we have fallen short. Exploration has become tainted with commerce and has lost sight of the improvement of man."

Temple's essay prompted a flurry of scrutiny by emerging intellects mostly in their late 20s to mid-30s. The wunderkid William Wotton quickly rebuked his claims in an essay published in 1694, and a young Richard Bentley further refuted them in a 1697 dissertation.

Swift, roughly 30 and by then ordained as an Anglican priest, and Charles Boyle, still an undergraduate, rushed to Temple's defense. In rebuttles released before Halley set sail, they took Bentley on directly. Swift's *Full and True Account of the Battle Fought Last Friday, Between the Ancient and the Modern Books in St. James Library,* completed in 1697, wouldn't be published until 1704, after Temple's death. A second collaborative work with Boyle would appear in print in 1698. The London satirist Alexander Pope, who was often derided by his critics as a "hunchbacked toad," would continue the defense of literary classicism, in his *An Essay on Criticism,* which included the infamous line "a little learning is a dangerous thing." It derived standards of taste from the natural order and was published in 1711, when Pope was 23.

Soon after the Battle of the Books erupted, a related conflict between science and religion surged. This time the skirmish was fought over Earth's history. Natural forces like gravity and laws of motion—now explained by Newton—could be applied to creation, the deluge, and the pending apocalypse or consumption of Earth by fire that was believed to await humankind. Thomas Burnet, master of the charterhouse, published the work that provoked this parallel contro-

versy in the late 1680s. It was probably the boldest endeavor to reconcile the competing accounts of religion and science. In the work, titled
Telluris Theoria Sacra, Burnet denounced the notion of divine intervention. He posited that, instead, Noah's flood was transpired by a
series of inevitable physical phenomena.

Halley became associated with this school of thought, though
perhaps unjustly, thanks to his discussions on the deluge which he
would publish many years later. The Royal Society types denounced
Burnet's take on science but sanctioned his ideas about natural law's
role in the machinations of divine will. By their logic, God was more
likely to function through immutable physical laws than like a fastidi-
ous watchman. Burnet would also go on to reexamine creationism
through the lens of such philosophy. Despite these distinctions, the
association would prove costly for Halley.

Swift, that feisty figure in the battle, would exercise his satire on
Halley and his colleagues even before he skewered science and politics in his well-known *Gulliver's Travels*. (On Gulliver's third voyage,
science is portrayed as futile unless used for human betterment.) Re-
ferring to Halley in a similar spirit, Swift wrote: "To him, we owe all
the observations on the *Parallax* of the *Pole-star*, and all the new *Theories* of the Deluge . . . Tide-Tables, for a Comet, that is to approximate
towards the Earth." His satire purports that the general public was
not interested in Halley's masterful application of theories of comets
or the specific intricacies in his tables. Taking on the whole Royal
Society, Swift said: "If *Scepsis Scientifica* comes to me, I will burn it for
a fustian piece of abominable curious virtuoso stuff."

Scepsis Scientifica, probably the best-known work of Joseph
Glanvill, defended the new "experimental philosophy," the promotion of which had been the focus of the Royal Society since 1660. The
work was subtitled *Confest Ignorance: The Way for Science*. Glanvill
argued that, without the Royal Society's pursuit of natural history,
"our hypotheses are but dreams and romances, and our science mere
conjecture and opinion." Even if this new science failed to explain

phenomena, he reckoned, it could advance agriculture, mining, and more. It is pleasant "to behold shifts, windings, and unexpected Caprichios of the distressed Nature, when pursued by a close and well-managed experiment," he wrote. Halley, in the thick of the battle, glibly countered Swift's attack. To him it was clear that moderns already exceeded the ancients. And astronomy was so much improved that "I had almost said Perfected."

In light of the difficult transition English society faced at the time, such a war of words full of satire smacks of folly. It seemed a superfluous waste of time when graver matters were at stake. Indeed, some scholars have argued that the war was actually more about the status of scholarship in society than about the clash between ancient and modern literature—the elitist scholar versus the layperson. No matter how complex its underlying nature, the controversy touched a public nerve. To be sure, the notion of a meshing of popularized science and religion was, at minimum, intriguing.

PERHAPS DUE IN PART TO HIS FAR-RANGING interests, Halley at age 29 was elected to serve the Royal Society's two honorary secretaries as clerk in 1685, more than a decade before his voyage. In this role, which required knowledge of Latin and at least a reading knowledge of French, he facilitated correspondence among the age's leading thinkers and was able to further explore questions that intrigued him. Besides corresponding with the world's elite on familiar topics of astronomy and mathematics, he discussed observation techniques with the Dutch microscopist Antonie van Leeuwenhoek and conversed with many other science greats about such nascent fields as geology, geography, physics, and engineering.

During much of this time, Halley also edited the *Philosophical Transactions of the Royal Society*, the society's banner publication first printed in 1665 and regularly since 1690 (making it today the oldest scientific journal in continuous publication). The scientific journal was in essence invented by the German-born Henry Oldenburg, one of the first two secretaries of the Royal Society. Data, experimental findings, and related theories now were widely circulated in a more timely fashion instead of being relegated to archives for compilation and study. Academies on the continent, such as the French Academie Royale, would also begin publishing similar periodicals.

At the time of Halley's voyage, however, the Royal Society was floundering. By 1698 the royally sanctioned coterie of intellectuals and dabblers had fallen on rough times despite its showing during the Battle of the Books. Its rolls had been plummeting for the past two decades. Although most of its illustrious members like Newton, Hooke, Wren, and Sloane remained, several of its founders, including Robert Boyle, had died. In the past three years the society had lost 60 members, shrinking to just over 110, less than half the number in 1670. Of these, roughly a third could be considered men of science. For the most part the ruling class gave it little financial support. For the past decade, nine different men had been elected to and held the office of president, only one of whose background could classify him as a scientist. That president was Sir Christopher Wren. And only about 10 (or 8.4 percent) of the fellows were members of the nobility and wealthy classes, down from 11 percent since the association's founding.

Halley had proved adept at keeping the society afloat during these lean years. At a minimum, he helped keep the society's key journal in print during this time. But he also skillfully moderated disputes that arose. With so many key questions unanswered in the world and institutions, personal beliefs, and priority claims under assault, it's not surprising that competing theories sparked bitter rivalries. Many such feuds were aired in the pages of *Philosophical Transactions*. The society called on Halley to intervene in the bitter skirmish between Danzig's doyen of astronomy Johannes Hevelius and the relentless Robert Hooke over telescopic sighting technology. Halley also tried his best to quell the almost calamitous bout between Newton and Hooke over who had first come up with one of the core ideas published in *Principia*. (Both battles will be detailed in later chapters.)

But Halley was not above throwing his own intellectual barbs either. He would eventually alienate his former teacher, England's first Astronomer Royal, John Flamsteed. But despite all this inner strife, many saw new hope for the Royal Society with the ascendancy of King William and Queen Mary in 1688.

IRONICALLY, HALLEY MIGHT NOT HAVE taken his adventurous sea voyage at all if not for troubles with that bastion of conservatism, the Church of England. Many thought he deserved the prestigious astronomy chair of his former professor, Edward Bernard, when the latter retired in 1691 as Savilian Professor of Astronomy at Oxford. Sir Henry Savile endowed the geometry and astronomy chairs at Oxford in 1619 to further mathematical and scientific thought. Even before his voyage, Halley, prominent from his work and association with Newton, seemed a natural choice. Both his alma mater, Queen's College, and the Royal Society submitted letters of praise. But there was a catch: The Church of England had to sanction his appointment as even Savile desired those holding the position to be "elected from among men of good character and reputable lives, out of any Christendom" as long as they had "in the first instance drawn the purer philosophy from the fountains of Aristole and Plato."

Religion dominated English life. And the church expected the staff of English universities, which were founded first and foremost to train the clergy, to practice religious orthodoxy. That had been the policy since passage of the Act of Uniformity in 1662, which mandated observance of the Articles of Religion of 1562 and added the requirement to the *Book of Common Prayer*. The teaching staff or fellows were mainly young men seeking eventual careers in the church, which forbade them to marry.

The church's dominance over the universities also hindered Newton's advancement at Cambridge. Despite his established piety, it is well documented that early on in his career Newton had trouble accepting the Holy Trinity as doctrine. Newton and Halley's religious stances, not helped by the largely "godless" first edition of the

Principia, are still fodder for academic debate. Newton added an argument to later editions that the physical world was created by an "Imperator," or otherworldy "Pancreator," but those editions didn't appear until the early 1700s, and rumors abounded that Newton was not a true Anglican, perhaps even a heretic.

Halley's problems were equally serious. Some powerful church men had branded him a skeptic, perhaps even an atheist. Specifically, they accused him of harboring unorthodox views based on some of his writings, including his introductory ode to Newton's *Principia.* Once their charges were levied, no matter their basis, Halley was responsible for proving himself innocent, which in the prevailing climate presented a seemingly insurmountable challenge.

Opposition from the Church of England also likely stemmed from Halley's public discussions in 1687 of one of the Old Testament's most dramatic stories: the biblical deluge. He irked the clergy by applying the latest science in gravity and magnetism to explain such sacred biblical dramas. Halley weighed in on an idea, originally proposed by Hooke, that rapid reversals of the north to south geomagnetic poles may have caused the epic flood that gave rise to Noah's Ark. Hooke suggested that such an event would cause the oceans to swell at the equator and flood the lands, as described in the Bible.

Halley expanded on the premise and got deeper into trouble. On the basis of observations made regarding minuscule shifts in measurements of latitude in Nuremberg over the course of the past few centuries, Halley argued that the shifts in the Earth's poles were extremely gradual. He contended that if Hooke's hypothesis were true, the flood must have occurred long before the time the religious experts estimated the world was created. He made the following introductory remarks to his paper, which he presented to the Royal Society:

> There have been many attempts made, and proposals offered, to ascertain from the appearances of nature what may have been the antiquity of this globe of earth, on which, by the evidence of sacred writ, mankind

has dwelt about 6000 years, or according to the Septuagint, above 9000. But whereas we are there told that the formation of man was the last act of the Creator, it is nowhere revealed in scripture how long the earth had existed before this last creation, nor how long those five days that preceded it may be . . . since we are elsewhere told that in respect of the Almighty a thousand years are as one day, being equally no part of eternity; nor can it well be conceived how those days should be to be understood of natural days, since they are mentioned as measures of time before the creation of the sun, which was not till the fourth day. And it is certain that Adam found the earth at his first production fully replenished with all sorts of other animals.

His words abraded many devout nerve endings. Yet even Halley knew better than to challenge biblical chronology too directly, so he proposed that something else—likely an external phenomenon like the impact of a comet or other astronomical body—must have triggered the flood. Or perhaps even the gravitational effects of a near collision sent Noah to his ark. This would also have helped explain the suddenness of such an event. Halley's crashing comet idea proved inaccurate, but his application of historical data to a scientific query proved a powerful contrivance.

In Halley's opinion, charges of heresy were levied against him only because the ideas in his paper were misconstrued. He had, in fact, credited comets with playing a dominant role in the divine plan in the paper he presented to the Royal Society in 1694. While he did also offer a theory on the deluge, it was hardly as radical as that of William Whiston, a mathematician, who would succeed Newton as Lucasion Professor at Cambridge. Whiston offered a detailed theory on a link between comets and creationism, the deluge, and even the apocalyptic millennium. So now comets, once thought to be harbingers of great catastrophes, had a new role as predetermined devices of the final judgment, in Whiston's view. The church also caught up with Whiston. Though a clergyman, he was eventually expelled from his professorship in 1710 for denying the doctrine of the Trinity—a change of belief influenced by Newton. For the most part, however, Newton took the secret of his own anti-Trinitarian views to the grave.

Halley himself thought that the account in Genesis omitted the collision's details not because of any oversight but merely because of the poor scientific comprehension of the day. So at the core, his views still conformed with the teachings of the church. In his view, as in Newton's, science could be the companion of religion.

Halley's critics, like Newton's, failed to appreciate such nuances. To them, his defense of his work was an affront to, not a validation of, the great Christian tradition of humankind's history and nature.

Due to the closed-door process of the screenings, Halley didn't know that the charges extended beyond his deluge theory. A powerful bishop named Edward Stillingfleet opposed his candidacy on the grounds that Halley was a "skeptic and banterer of religion." Halley met face to face with Stillingfleet, or at least his intermediary, Richard Bentley, in an effort to sway the bishop. According to one third-hand account, Halley told Stillingfleet that he "believed a God and that is all." Given his skills in diplomacy, it's unlikely Halley would make such an overtly contentious statement. No matter: Stillingfleet was unmoved by Halley's arguments. He believed the alleged skeptic was hiding his true sentiments. Others questioned Halley as well and were unsatisfied with his responses.

Historians find scant evidence supporting the allegations that Halley was an infidel. He was certainly a Christian, though an unorthodox one, it is agreed. Two letters penned in 1686 during his first year as clerk of the Royal Society, one in English and one in Latin, seem to back up the assertion. In the letters he details a "calico shirt brought from India woven without a seam, all in one piece" and writes that the shirt explains the scriptural mention "of our Savior's coat which was without seam."

Evidently the letters were fulfilling one of Halley's responsibilities as clerk: reporting on odd phenomena. (He also filed correspondence on such topics as a child with six fingers and six toes and a 37-year-old "little Man" in France who stood 16 inches tall and was practically dwarfed by his beard.)

Halley couldn't foresee that another observation made during his college years would later be used to slander him. Apparently while charting stars on the remote Atlantic island of St. Helena, Halley made the acquaintance of a couple who were expecting a child. Curiously, to Halley, the expectant mother was 52, the father 55. He reported the odd pregnancy to the society. According to Royal Society fellow John Aubrey, an older contemporary of Halley, "There went over with him (amongst others) a woman and her husband who had no child in several years, before he came from the Island and the woman was brought to bed of a child."

Astronomer Royal John Flamsteed spread rumors not only that Halley fathered the child of the 52-year-old woman on St. Helena but also that he seduced Hevelius's wife, Elizabeth, on a visit there in 1679. Some scholars suggest that Flamsteed, 10 years Halley's senior, who had once helped advance Halley's career, became threatened by his achievements and may have been the one pulling the strings behind the scenes to thwart Halley's candidacy.

In a letter to Newton at the time about Halley's candidacy for the Savilian professorship, Flamsteed wrote that Halley would "corrupt the youth of the University with his lewd discourse." In another letter to Newton several months later, Flamsteed continued his rant against his former protégé:

> If he wants employment for his time, he may go on with his sea projects, or square the superfices of *Cylindrick Ungulas* [part of a cylinder that resembles a horse's hoof]. He may find reasons for the change of the variation, or give us a true account of all his St. Helena exploits, and that he had better do it than buffoon those to the Society to whom he has been more obliged than he dares acknowledge. That he has more of mine in his hands already than he will either own or restore and that I have not esteem of a man who has lost his reputation both for skill, candor, and ingenuity by silly tricks, ingratitude, and foolish prate. Yet I value not all or any of the shams of him and his Infidel companions being very well satisfied that if Xt [Christ] and his Apostles were to walk again upon earth, they should not escape free for the calumnies of their venomous tongues.

"But I hate his ill manners, not the man," Flamsteed conceded. "Were he either honest or but civil there is none in whose company I could rather desire to be." Even Flamsteed wouldn't have stooped so low as to brand Halley himself an infidel, as one scholar points out. Newton never would have maintained a relationship with Halley if that charge were the least bit true. In the end, however, the Oxford chair was awarded to a less controversial Scotsman, David Gregory. Unlike Halley, Gregory was a true disciple of Newton, beholden to his patronage; Halley was more his peer than his protégé in the complex hierarchy of London society.

In time the writings of Halley, Burnet, Hooke, and others on the great deluge led to a new approach to evaluate biblical evidence using rational analysis. But Halley wouldn't freely discuss such views publicly for several decades. He wouldn't actually feel at liberty to publish "Some Considerations About the Cause of the Universal Deluge" until the controversy with the Church of England was long over. The issue of Halley's religious convictions remains unresolved.

Unfazed by the seemingly egregious slight, Halley was content to move ahead. He had won enough influence in and about London to realize an ambitious mission that on its face seemed next to impossible. He hoped to make an irrefutable intellectual contribution by service at sea. In many ways the long voyage would sate his adventurous appetite for knowledge better than any cozy university perch, and the isolation of a ship might offer some welcome tranquility. Whether findings from his *Paramore* voyage would tilt any key arguments of the day remained an open question as his sails finally caught a breeze out of London.

Halley sailed with the controversy sparked by the Battle of the Books still resonating in his mind. As the king's confidant, Temple had said:

> One great difference must be confessed between the ancient and modern learning; theirs led them to a sense and acknowledgement of their own ignorance, the imbecility of human understanding, the incomprehen-

sion even of things about us, as well as those above us.... Ours leads us to presumption, and vain ostentation of the little we have learned, and makes us think we do, or shall know, not only all natural, but even what we call supernatural things; all in the heavens, as well as upon earth; more than all mortal men have known before our age; and shall know in time as much as angels.

So Halley's quest had become something more than an inaugural science voyage. It was also part of the struggle to boost humankind's faith in itself. If Halley hoped his wanderings aboard the *Paramore* might produce tangible benefits for mariners, he also perhaps hoped it would help sway public sentiment in the never-ending War of Words.

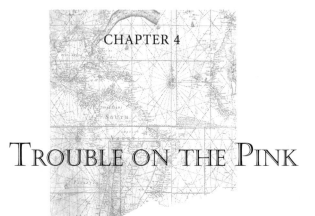

CHAPTER 4

TROUBLE ON THE PINK

O nce past Africa's northern coast, Halley thought he'd left the reputedly fiercest marauders behind him. Yet as the sails of the *Gloucester, Falmouth, Dunkirk,* and *Lynn* frigates blended into the seascape, the realization set in. As much as Admiral Benbow might have applauded Halley's quest, his fleet could ensure the *Paramore* safe passage only to Madeira. Naval duties and vibrant trade winds were calling his armada to the West Indies. A mere two and a half weeks after leaving the English Channel, Halley and his crew were adrift alone. It was four days before Christmas 1698.

Halley had much to ponder during the long hours at sea that would unfold before him. His attention was divided between captaining and plotting his experiments. Of course, his second in command was there to back him up—or so he hoped.

He had little or no idea what threats were coming up from the horizon. A good captain exuded confidence even when surviving seamen's journals prompted paranoia. At the time of Halley's venture, treacherous weather was blamed for about half of all shipwrecks.

After that a vessel's lack of seaworthiness and a captain and his crew's ineptitude were tied for the second leading cause of lost ships. Even with its obvious shortcomings, Halley's pink was in better shape than many a floating mausoleum out there pressed into service beyond its years. Earlier in his voyage, Halley informed the secretary to the Admiralty in London, Josiah Burchett, that "the Pink proves an excellent Sea boat in bad Weather." Other captains were more harshly tasked. The *Lindsey,* for example, was destined to be lost returning from Jamaica that very year. "She was old and crazy and not fit to go to sea."

Damage from shipwrecks was not tallied at the time of Halley's mission. Anecdotal evidence suggests all hands were saved in a majority of wrecks, maybe as much as 75 percent of the time. Many ships sank slowly or grounded near shore, enabling the escape of all on board.

Much of Halley's and the world's interest in magnetism was spurred by the drive to improve navigation. As commerce routes grew, in large part due to the burgeoning slave trade, fewer and fewer transatlantic voyages traced straight lines of latitude on paths that ran strictly east-west or west-east. Halley believed his ideas about Earth's magnetism, first documented six years prior in 1692, might help solve the nagging problem of how to determine longitude at sea on such courses. Its frequent misjudgment was blamed for many a disaster on the waves. The sheer number of lost ships, crew and cargo, and associated horrors warranted a search for better navigation techniques. And with the queen's blessing, Halley hoped to improve the odds of survival of even the most incompetent of captains. The Royal Society called on the maritime community to submit its sea observations and take more readings at sea when possible.

IN THE EARLY DAYS OF OCEAN NAVIGATIONS, the key measurement was declination or magnetic variation. These essentially interchangeable terms refer to the angle between true north and the direction to which a compass needle points. Thus, keeping the compass in working order was all important. In the absence of more scientific methods, captains would punish sailors for having garlic or onion breath. Early seafarers believed the odor was pungent enough to demagnetize a ship's compass.

While Italian Christopher Columbus is often popularly credited with discovering the existence of magnetic variation, the Chinese knew of the phenomenon in the 12th century. It's clear that Portuguese captains had also been aware of it for two generations. They navigated using the Pole Star to follow lines of latitude and noticed the needle "north-easted and north-wested" from time to time. The first description of an azimuth compass, which best detected the variation, was actually published in 1514 by Joao de Losboa in his *Livro de Marinharia.*

Soon enough, Portuguese philosophers reasoned that a "true meridian" of no variation may exist. That is, the magnetic variation from true north was zero along this special meridian. Many pilots believed such a meridian ran through the Azores and Canaries. Famed chief pilot of the Portuguese India fleet John de Castro would disavow them of such notions. On a voyage to India begun in 1538, he recorded 43 values for variation—many of which wavered by five or six degrees east. He employed what was known as a "shadow instrument," a device developed by Pedro Nunez, the mathematical adviser to the king of Portugal, essentially the forerunner of the azimuth compass. It was a round metal plate with a graduated edge that was connected to a magnetic needle and hung in gimbals, with the plate set on the meridian and used in tandem with the astrolabe to measure altitudes of the Sun. Castro, accompanied by Nunez's brother, Dr. Lois Nunez, demonstrated that the variation did not follow a set pattern, as philosophers had predicted: There was no "true" meridian.

The potential value of knowing how the needle wandered or the amount of declination was widely recognized in the early 17th century. Good navigators knew that determining one's true bearings was critical for staying on a prescribed course, especially one that traversed the Atlantic on a diagonal trajectory.

But the puzzle was more complicated than such enterprising pilots first realized. As early as 1633, Henry Gellibrand, chair of astronomy at London's Gresham College, demonstrated that the city's magnetic declination had shifted. Other observations revealed that similar variations occurred around the globe without any obvious pattern. Within two years, Gellibrand had recognized that "the variation is accompanied with a variation," that is, magnetic declination changes with time or experiences a secular variation. In other words, declination not only varied from place to place, but the lines of magnetic variation gradually moved at minimum from year to year—in fact, as much as a few minutes of an arc, analyses would reveal.

Halley was not the first Englishman to believe that magnetic variation could be used to find longitude. As soon as Gellibrand revealed secular change, a navigation teacher named Henry Bond, driven by substantial personal debt, began an investigation of declination. Bond boldy wrote a book, titled *The Longitude Found*, in which he claimed exactly that. He surmised that the variation would be zero in London but would gradually climb to the west. As this was true, Bond's ideas were given some credibility. In 1674, King Charles II established a committee of six, which included Robert Hooke and John Flamsteed, to explore Bond's method. Its creation was prompted by Le Sieur de St. Pierre, a Frenchman and alleged imposter, who lay claim to an impracticable rival method, based on lunar and stellar distances, and sought reward from Charles. Although the committee's members knew Bond's work itself was near rubbish, they sanctioned it on March 3, 1675, and recommended Bond be remunerated in order to dispatch St. Pierre. On the basis of the commission's full report, Charles immediately founded the Greenwich Observatory to make the lunar-distance method usable and placed Flamsteed in charge. (Some 150 years would pass before enough data were collected to do so.)

Over time many seafarers came to associate changes in magnetic declination with various destinations. For example, some sea captains preferred using changes in magnetic declination over conventional sounding as markers of location. William Dampier, a contemporary of Halley's who captained a Royal Navy ship a year after him, noted in one of his journals that 50 to 60 leagues off the Cape of Good Hope, declination was superior to soundings.

Astute sea captains knew they needed to be constantly compensating for such changes in the Earth's magnetic field. If a navigator made an incorrect allowance for local magnetic declination, a ship could quickly and unknowingly sail off course. Both under- and over-estimations of the local differences between magnetic and geographic north could cause a pilot to mistake a ship's true position. Exactly how mariners adjusted their compasses on the basis of magnetic readings is unclear to historians of science. Yet the compass remained invaluable. In the 1670s, English astrologer Henry Coley summed up the compass's use well: "Although it pretend[s] uncertainty, yet it proveth to be one of the greatest helps the seaman have."

At the time the *Paramore* sailed, a central question remained unanswered: How could magnetism vary if the Earth was a permanent magnet, Halley and others wondered?

HALLEY PLANNED TO SAIL to St. Helena, an isolated volcanic island in the middle of the South Atlantic, a no-man's-land for Atlantic voyagers, on the first leg of his mission. He'd sail past Madeira, the Canary Islands, and Cape Verde on the way. This southernmost province in the British kingdom had only in recent decades come under control of the English East India Company through a charter from King Charles II. Since its discovery 200 years earlier by Portugal, it had become established as a welcome pitstop on the sail to and from the East Indies. In 1672, when Charles declared war on Holland, the Dutch wrestled the strategic jewel away from the English and held it for a brief time. Fortressed by jagged cliffs, the little paradise, about a three-month sail from London, was familiar to Halley. Some 20 years earlier he had first made a name for himself there in the world of science and letters. Though under 50 square miles in size, the island offered good conditions for viewing the southern constellations—or so Halley believed when he selected the location.

A mere two and a half months into the *Paramore's* journey, Halley's log had grown thick with observations of all sorts. On the way to St. Helena, he wrote that the pink "[passed] through a Streak of Water in appearance turbid. But when in it we took up some Water, and it was full of Small transparent globules, but less than white peas interspersed with very small blackish Specks. These globules were so numerous as to make the Sea of a yellow muddy Colour. Their Substance appeared like that of our Squid . . . and there were two or three sorts of them." The yellow hue was an armada of jellyfish—an undocumented species at the time—according to one Halley chronicler named Dalrymple.

Halley had been recording his observations daily since sailing out of the Thames from Deptford. These included variations of the compass, wind, course, mileage, latitude and longitude from London, biological phenomena, and more. Despite the challenges of his ship and the open waters, he religiously took his research measurements no matter the conditions.

Halley intended to measure the magnetic variation over the vast test bed of the Atlantic and beyond in order to develop his hypothesis on the origin of Earth's magnetic field and understand how it changes. He hoped the results might solve the longitude problem once and for all. If the phenomenon proved to follow a regular pattern, he reasoned, it would be suitable for determining longitude at any point at sea.

Whenever the skies and seas cooperated and Halley was allowed a clear, steady glimpse of the celestial clockworks, he determined his position and calculated true north, conveniently lodged under that beacon for mariners, the North Star. True north was the direction toward the geographic North Pole, where the lines of longitude on a globe converge.

Halley used the two steering compasses couched within his ship to maintain his course. Comprised of little more than an iron wire attached to a marked card and encased in water-tight glass, the devices hung inside a square wooden box or binnacle on gimbals to keep them level against the ship's motion. Navigators have long exploited the compass needle's tendency to seek the north. The devices are built so that their magnetic needles can rotate freely in the horizontal plane. But as one approaches the magnetic poles, the needle veers more and more from the horizontal. At the north geomagnetic pole, the needle points down, while it points up at the south geomagnetic pole.

To measure magnetic declination, Halley relied on two bearing or observational compasses. At periodic intervals he eyed their quivering needles and recorded as best he could the movement away from magnetic north. Just how much it was off kilter from true north depended on where the ship was when he took the reading. (Since his school days Halley had been well aware that the horizontal angle between the direction of true north and magnetic north varied depending on geographical location.)

Compasses had a privileged position on most ships but especially on the *Paramore*. Halley's top officers, especially Harrison, were no strangers to the inner workings and importance of the compass. In his book on longitude, Harrison wrote: "Suffer no great guns or other iron too near your compasses." He had included the old mariner's maxim to help describe the problem of compass deviation, wherein the needle improperly deflects due to nearby iron.

UNDERSTANDING OF THE EARTH'S MAGNETISM had evolved gradually. Halley's contribution would fold into a history involving many countries across centuries. Although the compass likely developed independently in many regions, including China, Arabia, and Greece, the Chinese were the first to write about the compass for navigation in 250 B.C. It was described in the works of Hanfucious. Historians of science believe the Chinese were using the compass for navigation at least by the 14th century.

The first compasses were made of magnetized magnetite. The black, dense iron compound was called lodestone for "leading stone." The word "magnet" most likely derived from a place in ancient

Macedonia called Magnesia, where lodestone deposits abounded. Ancient Greeks also knew of the mysterious properties of lodestone.

The whole notion of a north magnetic pole originated with European navigators' need to understand the compass's directional properties. The compass had been exported from China to Europe by the 12th century, although it wasn't used there—as in China—for navigation until centuries later. Oddly enough, the Chinese viewed the compass as a southward-pointing device. This change in orientation would be key to devising theories about the nature of magnetic poles.

Incredibly, the sea compass technology remained quite primitive throughout the 17th century. Standard compass needles, commonly made of hard steel wire, did not retain their magnetism for long. The craftsmanship was often rudimentary. For example, the needles were frequently improperly attached to the card, skewing their readings. Still, valiant explorers of the 15th, 16th, and 17th centuries somehow successfully reached their destinations.

When four of Admiral Sir Clowdisley Shovell's fleet were lost on the rocks west of Scilly in October 1707, reducing England's military might by a couple thousand able men overnight, the Navy investigated the state of its compasses on its ships. Only 3 out of 145 ported aboard ships functioned properly! The majority of the compasses were floated in wooden bowls. Despite the fact that brass bowls had been found to be superior, they were expensive and only issued to flagships or ships voyaging abroad like Halley's.

Distances at sea were measured by throwing a log overboard and then observing how much time passed before the stationary log played out the distance on a line of a certain length that was demarcated with knots. The practice was developed in the 16th century, and to this day, as in Halley's time, the speed a ship travels is measured in knots, which along with other vital data is recorded in a "logbook."

Dead reckoning navigation employed both the compass and the log to determine a ship's position. The compass indicated the direction and the log gave the speed through the water, but the inefficiencies of these instruments often yielded erratic headings.

Halley typically used several methods to determine magnetic variation. They all hinged on the notion that the position of a celestial object at a given time would give the direction of geographic north. Then he could compare the readings with the compass needle's position. The difference was the declination. The simplest was a double-amplitude observation. He would compare the local meridian, which is half the angle between the Sun or a given star's rise and set, to the needle's orientation. Yet on many a day at sea, taking a single measure of the amplitude of the arch of the horizon was more practical. It required one observation, say at sunset, but more calculation. Essentially, Halley took the difference between the Sun's true orientation and magnetic direction on the horizon relative to east or west as pointed out by the ship's compass. He typically then allowed for refraction on the horizon and then compared the result with the Sun's computed azimuth from geographical north.

Sometimes, however, Halley found a third method for measuring the variation more reliable. For this approach he used one of the two portable azimuth compasses, which he himself brought aboard the ship. Like the amplitude compass, they could be set up virtually anywhere on deck atop a tripod or stool. With them, the idea was to take a precise magnetic bearing of the Sun or another object to compare with its true bearing. Since the Sun can't be viewed directly, its orientation was determined by the shadow cast through a slit in the device. Halley then used spherical trigonometry to calculate the true orientation of the point on the horizon at which the Sun sets and rises given the latitude and solar declination. The variation could then be found by simple subtraction: It was half the difference between these two amplitudes. At sunset and sunrise, Halley would take two readings of the angle between the magnetic needle and the Sun. At these times the Sun rests on the horizon, making the angles between it and the needle easier to observe. The variation could then be applied to the position of the *Paramore* at midnight or basically any time of day.

If the Sun's altitude and time of the observation were also known, the azimuth of the Sun at that moment could be calculated directly instead of taken from tables carried aboard the ship. The method required that the local time aboard the ship be known and that the observer was a relatively skilled mathematician.

At the time of Halley's voyage, the azimuth compass had not yet been accepted over the amplitude measurements as a better method to take local magnetic declination, however. Most navigators weren't yet knowledgeable enough in spherical trigonometry. Halley would take azimuth readings as backup: "Finding it scarce possible to get an amplitude in this cloud and foggy climate, I am forced to take the Sun's azimuth when it is low," he once noted in his log.

Halley's self-taught contemporary, Dampier, an ex-buccaneer and admittedly not a scientist, explained how variation all out stumped him from time to time:

> These things, I confess did puzzle me: neither was I fully satisfied as to the exactness of the taking the variation at sea: for in a great sea, which we often meet with, the compass will traverse with the motion of the ship; besides the ship may and will deviate somewhat in steering, even by the best Helmsmen; And then when you come to take an Azimuth, there is often some difference between him that looks at the compass and the man that takes the altitude height of the Sun; and a small error in each, if the error of both should be one way, will make it wide of any great exactness. But what was most shocking to me, I found that the variation did not always increase or decrease in proportion to the degrees of longitude East or West, as I had a notion they might.

The first real improvement in the azimuth compass was more than a century away.

Halley's first mate, Harrison, independently measured the variation as well, probably hoping his data would vindicate some of his assertions about a geomagnetic scheme for longitude, which had been summarily dismissed by the Royal Society before the voyage. The fact that his book *Idea Longitudinis* came out in 1696, four years after Halley first posited his hypothesis on magnetism, didn't help his book's acceptance. Halley and others had reported to the Admiralty that Harrison's book was more of a synopsis of ideas already out there and that it contained nothing novel. But some of his ideas for navigational practice were noteworthy. For one thing, Harrison had advocated that mariners take multiple readings: "Trust not to one observation, when you can have the medium of 5 or 6 or more, nor to one amplitude when you may have the mean of 3 or 4 azimuths," he wrote.

Both men shared the belief that longitude could be determined but with a painstaking amount of time and research. Harrison expressed it well in his *Idea Longitudinus*: "Some Blockheads are apt to say, the Longitude cannot be found; no, no it cannot Accidentally . . . but by Care, Diligence and Industry; it may be found, without which it cannot be understood."

WHAT ENGLISH SAILORS FEARED most, even above wrecking at sea, was capture or worse at the hands of pirates or Barbary corsairs. The threat probably lingered with Halley's crew even then. During wartime, French and Dutch privateers often collected crew as bounty as well. Historically, the corsairs, based in the North and West African regions of Algiers, Tunis, Tripoli, Morocco, and most commonly Sallee, were the most dreaded raiders. They enslaved their prey. Some of the wealthier captives were returned for ransom, but the average detainee faced a life of servitude.

Moslem raids on Christian ships had declined sharply from their peak 60 to 80 years earlier when 350 English subjects a year were captured by Barbary corsairs, scholars estimate. When Halley set sail, protective treaties with many of the marauding nations such as Algiers had been signed, but they were unenforceable. Random hijacks at sea continued. Moreover, not all African countries participated in even partial truces with England. Roving pirates from Morocco and Sallee continued to seize English ships though in much smaller numbers. By the 1690s, a dozen or so of such pillaging ships still roamed the sea in a given year—down from hundreds in the corsairs' heyday.

Meanwhile, the threat from European pirates known as buccaneers still loomed large. They had emerged in the pirating gambit in the 1660s and gradually expanded their territory from the Caribbean to the Indian Ocean. French privateers, though considered more benign than buccaneers or corsairs, were also to be respected. Between 1695 and 1713, they captured roughly 10,000 ships, a majority—perhaps 400 a year—of them English.

Torture of captives was prevalent. "It is, and ever was usual and common amongst privateers of all nations upon taking of their enemies, where they suspect any hidden treasure to be aboard, to use some torture or other," Halley's contemporary privateer Bartholomew Biggs said in a deposition in 1706 before the High Court of the Admiralty. Sticking burning matches between the bound fingers of a captive was the most common method of torture, he said, revealing practices from his 50-year career. Other techniques included crunching fingers in carpentry vices or "woolding"—supposedly the preferred practice of buccaneers—whereby the raiders slowly tightened a rope around a hostage's head sometimes to the point where his eyes popped out.

In most raids torture wasn't necessary. Sheer might or mere intimidation sufficed. And sometimes the greatest loot was food and water, priceless commodities miles from land. As we shall see, ample threat also existed from those humans closest to oneself aboard ship.

HALLEY WAS 20 WHEN HE FIRST visited St. Helena, traveling aboard the sailing vessel *Unity*, owned by the powerful East India Company. At the time, the island was the only territory in the Southern Hemisphere in England's possession. It is actually the summit of a composite volcano eroded by the elements and time. While the island itself is unappreciable in size, the base of the volcano is massive, extending to a depth of 12,600 feet. Overall, modern volcanologists put its volume at 20 times that of Mt. Etna, Europe's largest volcano.

Halley ended his schooling as a commoner at Oxford University to undertake his premiere adventure with his friend and assistant known only as Mr. James Clark. The trip served as Halley's primary schooling in life at sea. Aboard *Unity* he first learned the daily rigors of seamanship and glimpsed firsthand the obligations of a sea master, his first mate, and his crew.

As an Oxford undergraduate at Queen's College, Halley probably engaged in all the traditional coursework to continue his studies in Latin, Greek, likely Hebrew, and mathematics, which then included the spectrum of advanced work in geometry, algebra, navigation, and astronomy. Spherical trigonometry, which would come in especially handy in his later work, was also de rigueur. His dons included the Reverend John Wallis, Savilian Professor of Geometry and a founding Royal Society fellow, and his student Edward Bernard, who became Savilian Professor of Astronomy in 1673 after Wren. Scholars are unsure which of Halley's tutors would have had the most influence over his work.

Halley was most passionate about astronomy, for he claimed it had grabbed "hold of him early and would not let him go." Even before entering Oxford, he had earned a reputation as a shrewd star watcher. As a prominent London book and map seller named Joseph Moxon commented, according to John Aubrey's writings, "If a star were misplaced in the globe he would presently find it." By this time Halley was also possessed by geomagnetism. His first known scientific observation, while still a 16-year-old pupil at St. Paul's School in London, was of Earth's magnetic field. In 1672, about a year before college, he measured the declination of the magnetic compass. (A decade later he published the observation in the Royal Society's flagship journal *Philosophical Transactions*.) He would devote his life to the pursuit of his wide-ranging passion to understand the physical relationships between Earth, the sea, sky, and beyond.

Not content just to read about the work of others in an Oxford classroom, Halley struck up a correspondence with England's top astronomer, John Flamsteed. Later, the two collaborated on numerous observations. Flamsteed was impressed with Halley's imagination, ambition, and powers of observation and in this period could only be proud and not jealous of his protégé.

During Halley's time at Oxford, Flamsteed had undertaken cata-
loging all the stars of the Northern Hemisphere. By employing the
latest observational techniques, he achieved a degree of precision 40
times greater than that of Tycho Brahe, the celebrated 16th-century
Danish astronomer who, before the invention of the telescope, logged
more than 1,000 stars and revolutionized gazing accuracy in his era.
Though equipped with the best instruments money could buy, the
wealthy Tycho had no telescopes or reliable clocks. He relied on the
motion of heavenly bodies for reference. Flamsteed's new positions
were accurate within 10 seconds of an arc. Instead of relying on open
sights, Flamsteed, like Halley and others, used instruments with telescopic sights that were attached to the frames of angle-measuring
instruments, such as the sextant or quadrant. Invented by Middleton
amateur astronomer William Gascoigne, the sights extended the
length of the radius of the instrument, on average about six feet. Although equipped with the latest built-in telescopic sighting technology, Flamsteed, who paid for most of his instruments out of his own
pocket, observed from Greenwich, where most of the Southern Hemisphere is not visible. Cassini and Hevelius were also documenting
northern stars in parallel efforts. Not intimidated in the slightest by
the old guard, Halley saw an opportunity to contribute.

With the support of Flamsteed and many others such as Hooke,
Wren, and his leading patron Sir Jonas Moore, Halley boldly hatched
a plan to observe the southern stars himself. For Halley, though he
had been excelling in his studies at Oxford, this side survey trip was a
risky undertaking to tackle before receiving his degree. Halley's father, "respected as a Gentleman of considerable Estate, and plenti-
fully happy in all the Goods of Fortune," according to a leaflet
circulated at the time of his death, agreed to finance the expedition at
a handsome level. He gave Halley an annual allowance triple that of
Flamsteed's salary as astronomer royal, or roughly 300 pounds, which
is equivalent to about 157,200 pounds today. Before leaving, Halley
garnered permission not only from the British government and King
Charles II but also from the mighty East India Company, which had
controlled St. Helena since 1651 and was his best possible transport

to the island. (Boyle may have supported the venture as not only a dynamic Royal Society fellow but also a director of the East India Company.)

Halley and Clark arrived on the isle in February 1677 and immediately set up shop on Diana Peak, a mountain in the center of the island that soars some 800 meters above sea level. Halley's observatory was about 100 yards above sea level on the mountain's northern slopes. From his fern-covered perch, Halley had a breathtaking view of the whole island, from its majestic volcanic crags to its thickly forested briars to its rare sandy bays. Some 150 years later, Napoleon's tomb would be among the sites visible from Diana's slopes.

A sextant and a quadrant, specifically built for the expedition, made up the core of Halley's observatory. His customized brass and steel sextant, essentially a metal frame extending 60 degrees from one side to the other connected by a curved scale, had a 5.5-foot radius and a built-in telescope mounted on one side of the frame to afford improved viewing accuracy. The device was used to measure the angle or celestial arc between two neighboring bodies. To determine the angular distance between two stars and thus their apparent positions, for instance, an observer fixed one side of the frame of the sextant on a given star and moved the other sight to line up with a second star. Halley's quadrant, a device that measures the altitude of the Sun from the horizon, was standard, the common 2-foot radius variety. He also brought along a 24-foot telescope that his father had given him when he embarked on his academic career and a pendulum clock.

Halley, of course, used the quadrant to maintain the correct time, a calculation critical for determining a star's position. Amid his travails he most curiously realized that the length of the pendulum on his clock required shortening to keep correct time on St. Helena, even though it worked perfectly back in London. He was baffled as to why. His future friend Isaac Newton would one-day follow-up this line of inquiry. The pendulum length needed to be changed because Earth is not perfectly spherical: It has a bulge at its equatorial region, within which St. Helena is located.

The far-flung island boasted less than 500 inhabitants at the time, nearly half of them imported slaves from the East Indies, Madagascar, and elsewhere. The rest of its denizens were mainly English farmers lured there by the company for the promise of 10 acres and a cow—20 acres and two cows if a man were married or betrothed. Finally, both Portuguese and Dutch settlers who first introduced slaves to the island filled out the population rolls. Since its discovery, voyagers to the Indies gladly watered there, especially on return voyages. About the time of Halley's visit, passing ships began to be required to pay a toll: one Madagascaran slave. For a visitor the cost of living was very high, but Halley's allowance was three times the annual compensation of the island's ruling governor.

During Halley's stint on St. Helena, the British monarchy replaced the governor, Richard Field, on the basis of reports of his "ill living" at the sizable company plantation. Among other things, he failed to treat the young astronomers with "all respect and kindness" as ordered. Apparently he lacked the political chops to successfully balance the whims of the East India Company's desire for profits and the needs of the islanders. The company feared his insolent demeanor might jeopardize the island's prospects. Yet Field's lack of graciousness as a host did not impact Halley's work much.

Although the island was often shrouded in clouds and mist, Halley and Clark plotted more than 340 stars over the course of the next year. Cleverly, Halley took pains to ensure the longevity of his masterly observations. He surveyed some northern stars from St. Helena and the distance between various stars, so newer, more accurate observations of northern stars would not render his catalog obsolete. His expertly compiled chart could simply be recalibrated when necessary from the relative distances.

When the sky clouded over as it often would, Halley would spend time gazing closer to Earth where wild partridges and turkey hens roamed and a host of exotic flora and fauna thrived. The small island never bored him. He noted a curious plant that "bore perfect plants with a root on the extremities of its leaves and those sometimes will have others or grandchild plants." He described a polypus called a cuttfish by locals. The fish could walk on dry land like a long-legged spider. But when pursued, it could dash into the water and propel itself by a motion like respiration, emitting a dark russet-colored ink that turned the clear blue waters opaque. If that were not enough to elude a predator, "he will stick so fast to the rocks by means of several acetabula on his points or legs that he will be torn in pieces before he will let go of his hold." Islanders there also kept flounders for sale year-round in sea pens in the salt marshes, he reported. But the fish couldn't breed there because the muddy bottom destroyed their spawn. Such observations he'd later report to the Royal Society in its journal books—some he entered in his own hand when he assumed the job of clerk.

In November of that year Halley also made an important first observation in the heavens: a full transit of Mercury across the face of the Sun. Halley knew such an astronomical event could be used to measure the distance to the Sun. "I can say in truth," he said, "that if one can find the true position of [Mercury's] path with the Ecliptic, without difficulty my observations will provide the parallax of the Sun." Halley knew that from the parallax—the angular displacement—the distance of a celestial body could be determined. That is, if the relative position of the Sun's apparent path in the sky, or the ecliptic, is also known with respect to another body like Mercury, then the distance to the body can be calculated. Although Halley planned to observe the positions of the other planets, the weather did not cooperate with such ambitions. He mentioned his observation of Mercury's transit and celestial anomalies in a letter to Sir Jonas Moore, his lead patron who also helped found the Royal Observatory at Greenwich and personally financed many of its first instruments. Moore, surveyor general of the ordnance, worked from an office in the Tower of London and served as a communications agent to the Royal Society for Halley, Flamsteed, and others. In the Moore letter, Halley also noted that the Southern Hemisphere lacked a pole star like the North Star.

On returning to London in May 1678, Halley immediately sought to publish his astronomical findings and set his career in motion. He published a catalog of star positions, entitled the *Catalogue of Southern Stars*, and a celestial planisphere, or star chart. They were the first ever compiled from telescopic observations. He dedicated a constellation to Charles II, naming it *Rober Carolinum* or Charles's Oak. The work secured his M.A. from Oxford at age 22. Charles conferred the degree of master of arts *per literas regius* and "without performing any previous or subsequent exercises for the same" in November.

For the great minds of this period, self-directed study was not that unusual. Isaac Newton had gained most of his science education through independent study at Cambridge. John Wallis, a Cambridge student from 1632 through 1640, learned mathematics, "not as a formal study, but as a pleasing diversion, at spare hours; as books of Arithmetic or others mathematical fell occasionally . . . [his] way." Halley also followed in the footsteps of the likes of Jeremiah Horrocks, an up-and-coming astronomer at Cambridge in Wallis's day, and Wren at Oxford, who both contributed to the body of knowledge of modern science by publishing papers while still engaged in university studies.

Halley's southern star charts also conveyed prestige by securing him election as a fellow of the Royal Society and launching his reputation in the community as a man of scientific competence and verve. Astronomers and navigators alike made wide use of his charts. For Halley's efforts, Flamsteed paid him a high compliment, dubbing him the "Southern Tycho," after Tycho Brahe, who first cataloged the northern stars from Hveen, an island off the coast of Sweden.

As HALLEY VEERED OFF the western coast of Africa near Cape Verde in early January 1699 still heading to St. Helena, the *Paramore* sighted two ships in the distance. After some time, they were identified as English merchant ships.

Halley awaited a proper salute. None came. Before Halley could change heading, a thunderous crash erupted toward the rear of the hull. Off the port side, puffs of smoke emanated from the two frigates that were closing range. Great and small shots hurled toward the ship, crashing into the nearby sea.

The *Paramore* was outmanned and outgunned. Halley knew his pink didn't have a prayer of outrunning a merchant ship built for speed and that the odds of winning a fight were slim. He moved windward of the attackers, bracing the pink's headsails to the mast in surrender, still flying the Union colors. Likely somewhat agitated, he sent the dinghy over to inquire why a sister ship was firing on the research vessel.

The captain of the lead ship, the *New Exchange*, doubting the authenticity of its flag, had taken the *Paramore* for a pirate vessel and claimed to be "obliged to do what they did in their own defense." In fact, two masters aboard claimed that the *Paramore* was the very ship that had recently hijacked their former charges and released them of their commands. Perhaps relieved at the irony of the incident, Halley apparently accepted the misidentification at face value. That his pink emerged unscathed probably assuaged any urge for retribution. After all, pirates to deceive their victims commonly flew false flags of other countries, among other tactics. Moreover, nary a captain had seen a vessel like the *Paramore* charged with a scientific mission before. Halley quickly shrugged off the episode as other worries encroached on his sea laboratory. He set a course on January 6 for Trinidada (modern Trinidade) off the coast of Brazil.

THE ONLY UNIVERSAL LAW of the sea, according to lore, was that strange, inexplicable things happen. In the lingua franca of turn-of-the-century sailing, there was no such thing, no matter how benign the surface waters, as a safe harbor. Squalls emerged from dead calms, sailors deserted, mates disappeared overboard, captains ran amuck, entire crews became cannibals or mutinied for no apparent reason. The science of Halley's day was no match for such phenomena. Rather than the laws of physics, ordinary sailors often looked to the supernatural for an explanation, so much so that some 17th-century captains often hid their precious compasses from common sailors, wary their reactions might be unpredictable.

Superstitions abounded among many a mate. Some, like Halley's contemporary seaman Edward Barlow, a merchant captain who kept extensive journals, believed the wind to be the breath of God or that vengeful witches conjured wicked storms and uncooperative winds. Others swore ghosts of deceased sailors haunted their ships. Still others were plagued by mythical portents of death and disaster and lived in fear of great sea monsters like the "Pongo, a terrible monster, half tiger and half shark," as one historian tells. Most sea superstitions were more mundane, however. Sailors commonly linked the behavior of birds, fish, and other marine mammals, such as flocking of flamingoes or herding of porpoises about the ship, to weather patterns, as William Dampier described in captivating accounts of his voyages published at the turn of the 18th century, including *A New Voyage Round the World*, *Voyages and Descriptions*, and *A Voyage to New Holland*.

Sailors also turned to natural events like the appearance of shooting stars, comets, or rings around the Sun or Moon as harbingers of impending danger. For example, they believed the strange electrical discharge that sometimes appears near the top of the masts after a violent storm at sea—called St. Elmo's Fire—was a divine signal of sorts or even the embodiment of spirits. Per chance, some of these methods worked from time to time. The notion of using sea and sky phenomena to better manage a voyage was more in keeping with Halley's thinking than outlandish tales of Tinkerbell and other sea fairies guiding the way. If the *Paramore's* captain had witnessed such odd phenomena, he surely would have reported and investigated them; for although Halley had taken leave from his position as clerk of the Royal Society, which he'd held for 13 years, a large part of his job entailed documenting the discourse on perplexing issues.

As THE *Paramore* PRESSED on to the slave and sugarcane island of Trinidad, the ship's water supply was running low and occasional clouds of mosquitoes plagued the crew. Maintaining a steady reserve of fresh water was a chronic challenge for sea captains on long voyages. Although they had refilled all the water casks on the Isle of May, their progress since then had been slower than expected due to feeble winds. In fact, some days they traveled less than 20 miles. Caution called for rationing, and Halley limited each man to three pints of water a day.

Halley's pink was holding its own. The crew knew her faults well enough so they might seem mere quirks. When wind was slightly off the beam or somewhat forward of her, she could barely hold a forward tack due to her especially poor lateral resistance. When off the wind, she would easily clock a respectable speed and cover an easy 80 miles a day in good weather. However, in temperate breezes with her considerable forward resistance, she moved at a slower clip than a comparable merchant vessel. She wasn't all that large compared to the average seagoing vessel, but sizable enough that many tasks such as weighing anchor and getting under sail were major undertakings that involved all crew members.

Like most sea masters, Halley likely took the arduous daily work—the physical strenuousness and long hours—of sailors for granted, making little mention of it in his journal or ship's log. Halley sensed early on that his officers might not be up to the challenge of properly disciplining the crew. In a letter to the secretary of the Admiralty, Josiah Burchett, on leaving the English Channel, he expressed apprehension over "the weakness of his officers."

Sailors worked long hours with few chances for uninterrupted sleep. Typically, crews were divided into two watches, staggered at four-hour intervals, to keep the ship on point 24 hours a day. Shorter two-hour watches called "dog watches" were interspersed usually in the evening from 4 to 8 p.m. to rotate the watches throughout the day, so they didn't fall at the same time every day. In this way, men usually alternated between 10- and 14-hour days, though necessity often mandated longer workdays.

As captain, Halley probably wouldn't have routinely stood watch.

He would have left that chore to Lieutenant Harrison and to his boatswain. Yet as captain Halley was likely on duty around the clock. (His carpenter and cook probably didn't stand watch either.)

Maintenance of wooden ships like the *Paramore* was especially labor intensive, and when the crew wasn't actively engaged in sailing duties, there were lots of other things to do. Rigging, blocks, and cables required regular overhaul. Sails needed constant upkeep. Every piece of timber needed proper care to prevent waterlog. Ships' carpenters often enlisted help, scraping and repairing planks and sealing timber with caulk, tar, grease, and paint.

Most ships of the day chronically leaked, and the *Paramore* was no exception. Sailors were also charged with pumping out water that was taken on. Aboard the *Paramore* the task was perpetual.

Despite the arduous days, there is no evidence that Halley was an especially difficult taskmaster or that he abused his men in any way. To the contrary, Halley had taken efforts to ensure the health and success of his crew more than most captains. Of course, the paternalistic discipline aboard, especially that of the Royal Navy, was inherently strict. And its severity only increased during the Restoration period, scholars tell. The use of a whip at its peak to punish a court-martialed offender "rarely exceeded a hundred lashes," according to J. D. Davies. By any standard, England's naval code was harsh.

ONE MID-FEBRUARY MORNING some time after 2 a.m., Halley awoke to observe the helmsman suspiciously steering off course. The westerly heading would mean the ship would miss his target: Fernando de Loronho, a nearer island than Trinidad, where Halley hoped to replenish the ship's ever-dwindling water supply.

The boatswain protested that his error was accidental. The crew contended that the candle in the binnacle, encasing the steering compasses, had gone out and could not be relit. But Halley knew better.

When confronted, his boatswain immediately set the ship back on course. Had the *Paramore* continued on this rogue path, it would have collided with a sunken rock waiting off the island's southwestern coast. If Halley suspected Harrison had something to do with his botswain's behavior, he took no immediate action. He did not let doubts about his crew show. But he became more vigilant now, perhaps paying as much attention to his crew as to his daily measurements.

When morning broke, the pink reached Fernando de Noronha and dropped anchor off the more benign leeward shores. Turtledoves cooed and land crabs scurried in abundance, as Halley described in his journal. The atoll was devoid of any hallmarks of human trespass such as feral goats and pigs. Much to Halley's chagrin, the stopover was a bust: no fresh water was to be found—only an abundant supply of wood. Halley instead used the break to do routine maintenance on the ship's hull, shrouds, and mast. He also sketched a map of the island in his log. Very likely the crew had definite thoughts about their leader's decision to linger there, more grist for Harrison's mill.

Halley was not as fluent in the colorful banter of sailors as was Harrison, a shortcoming he frankly acknowledged. A mixture of necessity and chaos had forged a sea vernacular comprised of mainly one-syllable words. For example, "by" meant sailing by the wind or on a bowline; "bear down" referred to sailing downwind rapidly toward a landmark; the "bitter end" was the loose, unsecured end of a line, often an anchor cable; and "pipe down" was the botswain's call for all hands to go below. Seamen could easily converse in this terse tongue at the peak of a storm or while under attack.

Although Halley spoke like a proper Englishman, his crew's disrespect was undeserved. He hadn't made any gross navigational gaffes in commanding his ship. In fact, Halley had actually shown up Harrison on several occasions with his piloting acumen and geographical knowledge. Exactly how the *Paramore* was steered into the doldrums remains an open question, of course. Perhaps it was purely bad luck or their squabbling over course plotting that took a toll.

Meanwhile, Halley's displeasure with his crew was mounting as winter came to the Southern Hemisphere—probably more so than most captains, given the stakes. His officers' resentment overshadowed any veneration they displayed for science. Yet in Halley's mind the outlook for the mission remained bright.

DANGEROUSLY LOW ON WATER, the *Paramore* pushed on for the northeastern coast of Brazil. They made landfall on February 26, one week later, and anchored off shore at Paraiba. On their arrival, the Portuguese governor of the region Dom Manuel Soares Albergaria sent his deputies to greet Halley. A sergeant major and an interpreter invited him to refill his empty water casks the next day from the Paraiba River. He also stocked up on tobacco and sugar. Afterward, Halley dined with the governor and impressed him with expensive gifts. Albergaria was "very obliging and civil," Halley reported to the Admiralty. But the Portuguese officials "were very willing to find pretenses to seize us and tempted us several times with a sort of wood they call Poo de Brasile which is an excellent dye, but prohibited to all foreigners under pain of confiscation of ship and goods." Keen to their ploys to entrap them by enticing them to purchase the contraband wood, Halley refused all trade with them. Once Halley had replenished his water provisions, he was ready to shove off as soon as possible.

During his sojourn at Paraiba, Halley measured longitude astronomically by observing the end of the Moon's eclipse as well as the magnetic variation using his compasses. He had hoped to survey the estuary, too, but the Portuguese officials, still untrusting, denied his request.

Departing from the Brazilian coast on March 12, Halley by now had given up on the notion of wintering there. He recorded in his journal soon after that "my Officers showing themselves uneasy and refractory, I this day [March 16] chose to bear away for Barbados in order to exchange them if I found a Flagg [a flagship] there." Whether due to his unruly crew, mistimed winds, or winter's onset in the South Atlantic, Halley's meandering return to St. Helena would be postponed. For the time being, he abandoned his plan to venture farther south, much to the pleasure of his impatient crew. He advised the Admiralty that instead he'd "adjust the Longitude of most of the Plantations [in the West Indies] and see what may be discovered in relation to the Variation of the Needle in the Northern Hemisphere."

Things had been relatively quiet. Then, while approaching Barbados, the final straw came from his top lieutenant. Harrison defied a direct order from Halley. He "pretended that we ought to go to Windward of the Island, and about the North end of it, whereas the Road is at the most Southerly part almost. He persisted in this Course, which was contrary to my orders given overnight, and to all sense and reason, till I came upon Deck, when he was so far from excusing, that he pretended to justify it; not without reflecting [offensive] language."

In front of other officers and deckhands, Harrison said Halley was "not only incapable to take charge of the Pink, but even a Longboat." As James Burney noted in his 1815 account of Halley's Atlantic voyage: "Respect for science, however, did not operate sufficiently strong on the Officers of Dr. [Halley's] or rather Captain Halley's ship, to prevent their taking offense at being put under the command of a man who had risen without going through the regular course of service with the Royal Navy."

Enough was enough. Halley's lenience expired. In a letter to Secretary Burchett, Halley complained, "For a long time [Harrison] made it his business to represent me, to the whole ship's company, as a person wholly unqualified for the command their Lordships have given me, and declaring he was sent on board here because their Lordships knew my insufficiency." From then on, Halley stifled any further mutinous acts. He placed Harrison under arrest and restricted him to his cabin that night and took charge of the ship. He reiterated his orders and made sure the crew complied. Under Halley's stern authority, the *Paramore* reached Barbados with surety.

Halley would later note the irony of Harrison's selection for the mission: "My dislike of my Warrant Officers made me Petition their Lordships that my Mate might have the Commission of Lieutenant, thereby the better to keep them in obedience, but with a quite contrary effect it has only served to animate him to attempt upon my Authority, and in order there to side with the said officers against me."

By this point, Halley had relinquished any intent of wintering in Brazil and pushing southward in the spring. He had lost all patience with his crew's antics. He had no success finding a flag officer, possibly Admiral Benbow, to reprimand Lieutenant Harrison in Barbados, Martinique, Antigua, or St. Christopher.

With Harrison secured in his cabin, a frustrated Halley aborted the mission and set sail to return to London. Reaching the homeport in July 1699, likely full of personal disappointment, Halley pledged to pursue an official court-martial against not only his lieutenant but also several crew members. He needed to salvage his bungled mission. After arriving in Plymouth in the English Channel, Halley complained bitterly of his personal ordeal with Lieutenant Harrison in another letter dated June 23 to Secretary Burchett:

> For the future I'd take the charge of the Shipp myself, to show him his mistake; and accordingly I have watched in his steed ever since, and brought the Shipp well home from near the banks of Newfound Land, without the least assistance from him. The many abuses of this nature I have received from him, has very sensibly touched me, and made my voyage very displeasing and uneasy to me, nor can I imagine the cause of it, having endeavoured all I could to oblige him, but in vain. I take it that he envies me my command and conveniences on board, disdaining to be under one that has not served in the fleet as long as himself.

Halley was confident the Admiralty would find that the behavior of his crew had been intolerable. After all, he had proven himself a worthy captain, returning to London without losing a single life, a rarity in those days. He requested a new crop of officers with the hope that he might proceed once again with his mission. But first, there was his case against Harrison and his officers to settle.

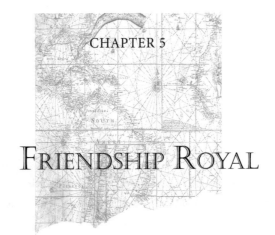

CHAPTER 5

FRIENDSHIP ROYAL

Back in London, gossip buzzed that Halley's crew had almost turned pirate on the mission. The speculation was fueled by the growing legend of Captain Kidd. The notorious one-time London privateer had became the "Scourge of the Indies" after his crew pillaged and plundered across the Indian Ocean. He was arrested in New York in June, a month before Halley's return. Kidd would soon be extradited to Europe's busiest city to stand trial by the Thames.

In Lieutenant Harrison's case, the rumors of pirating schemes aboard the *Paramore* seem to have been just that. But other allegations against him were serious enough that Halley would get his day in court. In response to a letter outlining Halley's charges, the Admiralty ordered a trial of Lieutenant Harrison.

The proceedings began July 3 aboard the *Swiftsure,* a frigate at port in the Downs. These calmer waters leading to the English Channel were sheltered by a series of sandbanks known as Goodwin Sands from the roiling waters of the North Sea. According to Pepys's diary, merchants seeking convoy to the Baltic would often congregate in the Downs.

It was an imposing sight. Admiral Sir Clowdisley Shovell presided. The *Swiftsure* was his ship. Nearly 50, Shovell was probably the most battle-savvy admiral of Halley's day. After serving as second in command during numerous successful attacks from Dunkirk to Calais, he was promoted to be Admiral of the Blue in October 1696. Out of three classes of flagship officer, the distinguished admiral, flying under a blue flag, held the rear of an engagement. Shovell returned from sea to serve in Parliament in 1698, the year of Halley's launch. He convoyed the king home from Holland in late October. In keeping with his prominent position, he sported a large emerald ring on his hand.

Three other admirals and eight captains comprised the formal court-martial panel. In addition to weighing Halley's claims against Harrison, they probed the warrant officers' actions, even though the Admiralty did not order it.

"I am sure that never any man was so used by a lieutenant as I have been, during the whole term of the voyage," Halley contended, apologizing for any offense to their lordships.

True mutiny was uncommon but still occurred in Halley's day. Sometimes it entailed bloodshed and takeover of a ship; other times it consisted of only passive defiance. Even in its tentative forms, mutiny was not taken lightly by the Royal Navy.

For his part, Lieutenant Harrison was probably lucky Halley hadn't found a flag officer in the West Indies. Men were hanged at sea for offenses far milder than Harrison's transgressions, but with no replacement at hand, Halley sought justice on shore. Away from land, admirals, in fact, had the power to have mutineers hanged on the spot—and they did. Death sentences could also be doled out for murder, desertion, striking an officer, cowardice, and aiding an enemy, that is, provided the fleet or a squadron's commander in chief concurred with the guilty finding of a court-martial by a minimum of five captains.

One curiosity in the annals of mutiny at sea, popularly known as the "round robin," emerged the year the *Paramore* set sail: It consisted of a piece of paper with two concentric circles scrawled on it. In the inner circle a group of sailors would "write what they have a mind to have done." Then they'd all sign their names outside the inner circle but within the outer circle. In this way all were equally guilty of the act, with no leaders and no followers but an "orbicular" chain of command. A "mutinous and seditious paper" was first passed the very year Halley sailed, according to a deposition, aboard a merchant ship, the *Fleet Frigott,* when her captain denied his crew permission to take shore leave on Tenerife. Soon enough, savvy captains sometimes averted mutinies by intercepting such a paper in circulation on their ship. Given its novelty, though, it is unlikely that such a device was passed around by Halley's crew.

LONDONERS AT LARGE ADORED a good hanging. And back in port mutineers were fair game—regardless of whether any semblance of justice was actually being served by the contemporary legal system. In the city, hanging "matches" were regularly held eight times a year—often enough to make the populace think twice before choosing crime as an avocation. These periodic civic holidays of sorts—albeit unofficial ones—could be considered the government-sanctioned blood sport of the age. By 1700, hundreds of crimes were in fact punishable by death—and the noose was the method du jour. Thousands of robbers, highwaymen, counterfeiters, and murderers swelled the rolls of the city's 150 prisons; the overload contributed to the evolution of a thriving underworld within the penal system. Fortunately for the condemned the system couldn't keep up with the rate of convictions; executions were implemented largely as deterrents to crime. Given the number of doomed inmates, excessive hangings would have sparked rebellion. Many felons managed to escape with their lives through bribes or sometimes by agreeing to be transported to the North American colonies or West Indies in exchange for pardons.

With deliberate pomp and splendor, London society glamorized the dread execution. The condemned were garbed in bright colors and fanciful garments perhaps more fitting for a church ritual. Ordered according to the severity of their crimes, they paraded their fleeting notoriety along the procession route from their prison cells at Newgate or elsewhere to the gallows at Tyburn, where the first permanent scaffolding of punishment was established for felons and spectators in 1571. It was named for the nearby brook, which emptied into the Thames.

Meanwhile, traitors of the state were executed at the more salient Tower of London. For high treason the sentence became more elaborate. After hanging by the neck to reach an appropriate shade of blue, the prisoner was cut down and disemboweled while still breathing. The executioners then hacked the body into four parts. Its disposal was left to the monarchy. Conventional wisdom held that the cadavers of the damned harbored mystical healing properties. After an execution, physicians and body snatchers alike competed for such spoils once darkness fell.

Pirates, mutineers, and other nautical criminals like Captain Kidd were hung at low tide at Execution Dock in Wapping on the banks of the Thames. Masses watched from floating barges. The procession was typically led, from the Southwark prison called Marshalsea to the shore via London Bridge, by an Admiralty marshal on horseback brandishing a silver oar. The prisoner followed atop a cart flanked by a prison chaplain. Along the journey, the ill fated were permitted to stop at taverns and imbibe to the point of inebriation—though the practice was usually not to the chaplain's liking. On reaching the Thames, as part of such rituals, the condemned were entitled to give a last speech. After the cart was pushed away, leaving the damned to dangle from the so-called fatal tree and the hangman's noose to brusquely tighten, tradition called for three tides to engulf the body of the executed before it was extricated from the gallows. The bodies of infamous pirates like Kidd were then tarred and hung in irons along the Thames as a warning to other would-be mutineers and pirates.

When it came to Lieutenant Harrison's fate, blood-thirsty Londoners would have to wait for another day. The court in fact found no evidence of mutiny aboard the *Paramore*:

> Under a strict examination into this matter the Court is of the opinion that Captain Halley has produced nothing to prove that the said Officers have at any time disobeyed or denied his command . . . though there may have been some grumbling among them as there is generally in small vessels under such circumstances and therefore the Court does acquit the said Lt. Harrison and the other Officers of his Majesty's Pink the *Paramour* of this matter giving them a severe reprimand for the same.

Indeed Lieutenant Harrison would not be coming even remotely close to a painful gallows death, or Tyburn's Tippet as it was called. Other factors probably weighed more heavily on ship. To begin with, it was unlikely that an officer like Admiral Shovell would rule against one of his own—a blueblood tarpaulin—in favor of a gentleman scientist.

Halley, believing he had overwhelmed the scales of evidence, was dumbfounded by the verdict. "Yesterday at the Court Martials I fully proved all that I had complained of against my Lieutenant and Officers," he asserted. "But the court insisting upon my proof of actual disobedience to command, which I had not charged them with, but only with abusive language and disrespect. They were pleased only to reprimand them, and in their report have tenderly styled the abuses I suffered from them to have been only some grumblings such as usually happen on board small ships."

In his mind Halley had skillfully and unequivocally revealed Harrison's motives. "My Lieutenant has now declared that I had signally disobliged him, in the character I gave their Lordships of his book, about 4 years since, which therefore, I know to be the cause of his spite and malice to me, and it was my very hard fortune to have him joined with me, with this prejudice against me." But such an impetus, Halley's cerebral critique of the lieutenant's longitude treatise, proved of little consequence to these practical men. Halley had failed to win the loyalty that tarpaulin captains often enjoyed. Seamen would often fight to their deaths for a leader they deemed worthy. Although Harrison essentially got off with a slap on the wrist and could have continued his career in the Royal Navy, he opted to rejoin the merchant navy soon afterward.

At least Halley escaped the fate of a like-minded intellectual explorer given a similar mission aboard the HMS *Roebuck* soon after Halley's voyage. When another naval outsider, William Dampier, was handed an expedition a year later to also explore for gold on *Terra Australis*, which he claimed to have sighted aboard a pirate ship, the *Cygnet*, in 1688 when it landed in New Holland. (New Holland and *Terra Australis* would prove one and the same and he would later be counted among the first British citizens to touch its mainland.) But trouble erupted as his ship reached Brazil. His lieutenant, George Fisher, also a tarpaulin, attempted to undermine his captain. Dampier clamped the mutinous lieutenant in irons and transferred him to a Portuguese prison, essentially leaving him to rot indefinitely until his transport could be arranged by the local governor.

Like Halley, Dampier would pursue a court-martial on his return to England in 1701. When an English convoy brought him and his crew back to London, Dampier was in for a surprise. After spending several weeks in the dingy Portuguese cell, Fisher had been released and launched a campaign against Dampier in London. This second civilian captain to sail for science and exploration under royal sanction would be sandbagged by the naval justice system just months after Halley had finished his second voyage. This time the Admiralty instead pursued three counts against the civilian captain and former buccaneer: the first for his treatment of Fisher; the second for allegedly contributing to the death of another crew member by "barbarous and inhuman usage"; and the third for losing his ailing ship, the *Roebuck,* though no men off Ascension Island. Perhaps biased by Dampier's pirate past, the Admiralty dismissed the mutiny charge against Fisher. They also ruled that the second charge was frivolous— a bogus claim by the grief-stricken wife of Dampier's boatswain who later died on another naval mission. But the Admiralty found him guilty of "cruel usage" of Fisher and insulted Dampier with a hefty fine: the sum was equal to three years' wages. The court found Dampier was "not a fit person to be employed as commander of any of Her Majesty's Ships."

Perhaps an even greater irony occurred later with the presiding officer of Halley's trial, Admiral Shovell. On a subsequent journey, Shovell hanged a sailor almost instantaneously for questioning his longitude calculations. But Shovell in fact was in error. On a fog-shrouded evening in October 1707, his fleet of 2,000 would ground off the Scilly Isles that extend from England's most southwestern shores. Although Shovell was one of two survivors to make it to the shore, a woman posing as a rescuer—a wrecker as such rather opportunistic beachcombers came to be called—suffocated him for that prized emerald adorning his finger.

BACK IN THE FAMILIAR RHYTHM of the London summer, Halley inevitably found himself among familiar haunts and friends. He set about sharing some early findings of his mission with the Royal Society. He made assorted presentations to his peers at the society's regular meetings. The elite club still had high hopes for the continuation of his expedition that fall.

Halley had already exceeded all expectations of the council of the Royal Society as clerk. Nearly 25 years had passed since Halley first met the illustrious city designers Christopher Wren and Robert Hooke when they went to Greenwich to survey the site for the Royal Observatory. And it had been 15 years since they had raised the question of how the planets orbit the Sun at a Royal Society meeting.

The trio's conversation had turned to the astronomical question on a January evening in 1684. At that time Halley was looking for a way to explain Kepler's laws of planetary motion, which involved not simple circular but elliptical orbits of the planets around the Sun. These elliptical orbits stipulated a mathematical ratio between a planet's time to complete one circuit and its average distance from the Sun. Kepler determined that ratio to be 2:3. But even he could not offer a physical explanation for it. One of Kepler's laws also stated that planets move faster as they approach the Sun on their elliptical orbits than when they move away from it and that these velocity changes are quantifiable.

Halley, like a handful of his colleagues, surmised that the force with which the Sun attracts a planet is dependent on the inverse square of the distance from the Sun to the planet, but he needed help devising the geometric proof. This equation is of no small importance because it hints at gravitational forces in the universe.

Meanwhile, Hooke had been thinking along similar lines and claimed to have solved the problem of the inverse square law. Wren was dubious. So he offered a prize to whoever could produce a proof in the next two months. The prize was a book worth 40 shillings.

When the deadline expired, no one, including Hooke, had demonstrated a satisfactory proof. "Hooke said he had it, but he would conceal it for some time that others trying and failing might know how to value it when he should make it public," Halley recalled in a June 1696 letter to Newton. "However, I remember Sir Christopher [Wren] was little satisfied that he could do it; and though Mr. Hooke then promised to show it to him, I do not find that in particular he has been as good as his word."

Halley would have to leave the question unanswered for a while. Before Halley could do much more to pursue it, he received some horrific news.

"A POOR BOY WALKING by the water-side upon some occasion spied the body of a man dead and stripped, with only his shoes and stockings on," according to a headline in an early March 1684 issue of the London *Gazette*. The details were more gruesome. His face and body parts were disfigured, one eye in particular. Identification had to be made from his footwear. The morning five days before the man disappeared he had cut out the linings of his shoes to make them fit more comfortably.

Little did Halley know the incident's profound ramifications. Although the cause of the man's death would never be truly known, Halley himself and his stepmother would never be entirely above suspicion. The naked corpse was that of Halley's father, Edmond Halley, Senior.

The verdict at the official inquest was murder. But nary a suspect was formally called. Some believed he was killed in conjunction with his role as a yeoman warder, or Beefeater. In addition to running his businesses, the senior Halley served as one of the uniformed guards who watched the prisoners and attended the gates of the Tower of London (a role he, as a prosperous merchant, fulfilled to exempt him from providing other services for the parish which he'd typically be obliged to pay at his own expense). The most important fortress in England, the tower then housed the Royal Mint and the Ordnance Office with its store of guns and military supplies. Still others suspected suicide. The fact that four socks were found on one foot and three on the other suggested "mental aberration," according to one account. Some supposed that Halley Senior had been distressed over the recent death in October 1672 of his youngest son, Humphrey, or possibly over financial worries, as significant rental properties had been destroyed in London's Great Fire in 1666. But money was probably not the source of his problems because even with the resulting temporary dip in rental income, he was by all measures a prosperous man.

In the grandest theory, the senior Halley's demise was linked to a Protestant conspiracy to assassinate Charles II and James Duke of York in April 1683 as they returned to London from Newmarket. The scheme, which became known as the Rye House Plot, was foiled, and two of the three chief conspirators were beheaded. But the third man, the Earl of Essex, was found dead in his jail cell at the Tower of London before he could stand trial. Whether the conspirator committed suicide or was himself murdered remains an open question. John Locke numbered among those who suspected murder. Some, believed to be connected with the original Rye House Plot, raised accusations.

In turn, they were tried and fined. But two other men were thought to know something about the death of the Earl of Essex. One was named Thomas Hawley and was a yeoman warder. Whether Halley's father was killed by accident because of mistaken identity or as the yeoman of the tower alleged to know how the Earl may have been murdered remains a mystery.

Halley was mum publicly about the matter. While his reaction to the news of the tragedy at times seems odd given his growing clout with the island nation's patrons and other elite, it's likely that Halley was merely treading dicey political waters. James II, a Roman Catholic convert, would assume the throne in 1685, the year after the tragedy.

In the decades to come, rumors would surface now and again that it might have been Halley himself who had a hand in his father's death. The sordid urban legend held that his "father went in fear of his life from Halley." Halley's father was worth roughly 4,000 pounds at the time of his death, which would be more than $4 million in U.S. currency today, but he left no will. Halley would end up battling his father's second wife, Joane, in court for control of the estate. (Halley's mother, Anne Robinson, had died when he was in his midteens.)

Despite his personal troubles, Halley would not let the inverse square question of elliptical orbits go unanswered. The query would propel him on a shorter but equally important journey to that of his *Paramore*—to visit Isaac Newton for the first time late that summer.

TRAVELING THE 60 MILES northeast from London to Cambridge's Trinity College was not easy before the metalled highways of today traversed the countryside, but noblemen did it all the time in order to conduct their affairs. What might have been running through Halley's mind as he watched the terrain roll by from his carriage is pure conjecture. Surely a part of him was still reeling from his father's death. The seemingly indefatigable Halley, then 28, was flush with ideas.

Fourteen years older than Halley, Newton shared his boundless curiosity and special interests in mathematics, comets, optics, and more. Among other things, the shy, skittish Newton by then had invented a new type of compact telescope, which used concave mirrors instead of lenses. The first in England, it earned him Royal Society fellowship. Although not the world's first, the Newtonian reflecting telescope offered several improvements over the period's refracting telescopes, which relied on a lens to focus starlight into an image magnified by the eyepiece. Not only did refracting scopes need to be very long in order to achieve high magnification, they also produced significant chromatic aberrations, bending different wavelengths of light differently like a prism. Newton's reflecting telescope used curved mirrors to focus the light by reflection, eliminating the chromatic irregularities, and didn't require as long of a focal length for comparable magnification.

Newton's related work also made him the focus of controversy. To merit his admission as a fellow, Newton had prepared a paper about light and colors. His earlier experiments had led him to theorize that white light was actually a rainbow mixture of all colors. The resulting hullabaloo over his controversial paper only increased his desire to be left alone. After his own public feud over optics in 1672, he resigned his Royal Society membership for a time and even ceased all correspondence. By the time of Halley's visit, Newton, in his 40s and his reputation established, had become a reclusive eccentric, engrossing himself in alchemy research after his mother died.

When Halley reached Cambridge University's enchanting medieval gates and massive elms, to his surprise, Newton told him that he had already solved the problem of planetary orbits, spurred in fact by communications with Hooke some six years earlier. As recounted by his close friend Abraham De Moivre, a French mathematician, "Sir Isaac replied immediately that [the shape of the orbital curve] would be an Ellipsis, the Doctor struck with joy & amazement asked him how he knew it, why saith he, I have calculated it." He saw it apparently as his own private puzzle, not bothering to publish this important mathematical proof. Although the temperamental thinker couldn't produce the proof right there and then for Halley (his papers were a mess), he promised to redo the proof of the inverse square law of planetary motion if he couldn't find the papers.

Soon enough, Newton redid the proof and sent nine pages titled *De Motu* to Halley in London. Impressed with the sophistication of Newton's reasoning, Halley visited him again at Trinity College inside his second-floor chamber that overlooked a tennis court. It was just off the Great Court, which was replete with a library, stables, fountains, and manicured lawns. He knew Newton was on to something big. The proof of Kepler's 2:3 ratio of elliptical orbits turned out to be just the tip of the iceberg, part of a broader revolution of the whole universe within Newton's head. But Newton again seemed indifferent and even more disinclined to publish his work, desiring to avoid the sniping feuds it would likely produce. "Philosophy is such an impertinently litigious lady," he said.

Ever congenial, Halley persisted. He coaxed Newton to elaborate his grand vision that might unify forces acting on Earth and in heaven in a book. Newton finally agreed to do so and immediately set to the writing. The work, in three parts, took him only 18 months, and it was his sole focus during this time. Halley would have to prod him along the way, pushing the asocial genius to finish the third book.

When a quarrel erupted with Hooke over his contribution to the work, Halley would intervene. Hooke had angered Newton when he publicly staked a claim to the "invention," saying Newton "had the notion from him" and deserved mention in the preface.

"I am heartily sorry that in this matter wherein all mankind ought to acknowledge their obligations to you," Halley scribed in a letter dated June 29, 1686, to Newton, that "you should meet with anything that should give you disquiet or that any disgust should make you think of desisting in your pretensions to a Lady whose favours you have so much reason to boast of. Tis not she but your rivals, envying your happiness that endeavor to disturb your quiet enjoyment. Which when you consider I hope you will see cause to alter your resolution of supporting your third book, there being nothing which you can have compiled therein which the learned world will not be concerned to have concealed. Those gentlemen at the society to whom I have communicated are very much troubled at it."

Halley not only had an intimate hand in crafting and publishing Newton's *Principia*, he personally financed its publication after the Royal Society rashly spent all its publishing budget on a history of fish. He took on the responsibility of shepherding the *Principia* through the press. He even forewent his salary as clerk—about 50 pounds a year—to help see this watershed science book through to publication. Halley had only been elected to the position in 1686, a year earlier. Back then a lower-middle-class family could live well on 50 pounds a year, which is the equivalent of $50,000 today. (Meanwhile, wealthy sea merchants grossed in the neighborhood of 300 pounds a year, sometimes amassing fortunes in the 10,000 to 15,000 pound range, or $10 million to $15 million today.)

In lieu of his salary, Halley was given 20 copies of the book, started by Francis Willoughby and finished by noted taxonomist John Ray, which had blown the Royal Society's budget, *Historia piscium* (or *The History of Fishes*). Each volume was worth roughly a pound. (Booksellers soon learned that scientific monographs were typically not best-sellers even in Halley's day.) Money, however, was never much of an issue for Halley. His father had left him with substantial passive income. He collected about 200 pounds a year in rent from the family's houses on Winchester Street alone. And there were other properties in the estate, despite the extensive loss of rental properties from the Great Fire in 1666.

Halley also ran interference between Hooke and Newton in their squabbles over the proof of the inverse square law. Initially, Newton considered acknowledging Hooke in the manuscript, but when news reached him about Hooke's credit-mongering behavior at a Royal Society meeting, Newton changed his mind, forever branding Hooke publicly, however unfairly, as "a man of strange unsociable temper." At one point Hooke so offended Newton that he might have stopped writing the work, suppressing the third book entirely, if not for Halley's aplomb and tact.

Many claim that but for Halley the *Principia* would not have existed. In later years Newton rather magnanimously referred to the *Principia* as "Halley's book." The volume spurred a revolution in thinking, disclosing for the first time basic laws of physics, "the unchanging order of things," that govern the universe.

According to Halley in his introduction to *Principia*, life simply would never be the same after this book. Newton's achievement, he said, surpassed the creation of cities, of laws that "curb Murder, Theft, and Adultery," of the making of bread or the intoxicating effects of wine or music. So many mysteries were now solved. Because of Newton, no longer would comets evoke imminent supernatural danger, at least not in the minds of educated people. The cycles of tides, the orbit of stars, the motions of objects, from a falling apple to the Earth's moon, were now explained, beautifully, simply, mathematically.

"Behold set out for you the pattern of the Heavens," said Halley, introducing Newton's work in 1687. "Laws which the all-producing Creator, when he was fashioning the first-beginnings of things, wished not to violate. . . ." Halley waxed as poetic as he could in these extracts (of a literal translation from the Latin) of his summary of Book Three of the *Principia:*

> Seated on his throne, the Sun commands all things to strive toward him by inclined descent; nor does he permit the starry chariots to move on a straight course through the immense void, but draws them, individually, into unchanging Orbits about himself as center. Now is revealed what the bending path of horrifying comets is; no longer do we marvel at the Appearances of the bearded Star. Herein we learn, at last, by what cause silver Phoebe proceeds with unequal steps; why, subdued thus far by no Astronomer, she refuses the bridle of numbers: why the Nodes recede and why the Apsides advance. We learn also with how much Force journeying Cynthia drives the ebbing sea, while with broken waves it abandons the seaweed and lays bare the sands suspected by Sailors; alternately beating upon the highest shores.

> Things which so often tormented the minds of ancient Sages, and which fruitlessly vex the Schools with raucous disputation, we perceive in our

path—the cloud dispelled by Mathematics. No longer does error oppress doubtful mankind with its darkness: the keenness of a sublime Intellect has allowed us to penetrate the dwellings of the Gods and to scale the heights of the Heavens.

"Now we are truly admitted as table-guests of the Gods," Halley continued. "We are allowed to examine the Laws of the high heavens; and now are exposed [to] the hidden strongholds of the secret Earth, and the unchanging order of things, and matters which have been concealed from the generations of past mankind." In short, there was no doubt in Halley's jubilant mind that this achievement exceeded all glories of past thought.

The collaboration culminated with the presentation of a 500-plus-page work called in full *Philosophia Naturalis Principia Mathematica* to King James II. Halley considered the work earth-shaking enough to present it to the king in person. And it was he and not Pepys, then a secretary of the Royal Society, who was afforded the honor.

Not many people, including the king, could understand it. The work was even over the head of most members of the Royal Society, including, by his own admission, Pepys. Supposedly, a student was overheard remarking to another after passing Newton on a Cambridge Street soon after *Principia*'s publication: "There goes the man that writ a book that neither he nor any body else understands."

In time the *Principia* would prove key to Halley's work in astronomy. Being so intimate with Newton's masterpiece could well have inspired the younger Halley to undertake his risky career-changing endeavor: the great magnetic survey aboard the *Paramore*. And it is likely that had *Principia* not created such an uplifted faith in scientific progress, Halley's expensive wayfaring mission to collect numbers around the globe might never have launched. Queen Mary approved Halley's mission about six years after the publication of *Principia*.

Despite the negative outcome of the court-martial, Halley was granted his request to sail again. He would be the only civilian ever again entrusted to command a royal ship after the debacle with Dampier and his crew came to a head in 1702.

On giving Halley approval in late July 1699 to take to the seas again, the Admiralty directed the Navy Board to examine the pink and either improve it or give Halley another ship. But Halley's favors with his patrons seemed to run out there. The board opted to fix his pink, not assign him a better vessel as he'd desired. Halley had vetted his new officers but had to reconcile himself again to the quirks of the *Paramore*.

SECOND VOYAGE:
1699–1700

CHAPTER 6

OUTWARD BOUND

I t was between three and four in the afternoon, according to the cant of the October sun. At sea for a month, the *Paramore* was approaching the Canary Islands. Although Halley's call for a new ship had been immediately denied back in July, his request for new officers was fulfilled. Only seven of the original crew returned. The monarchy had granted him 24 men—four extra because his botswain had only one arm. Halley received his new charge as master and commander in late August 1699 and his sailing orders in mid-September. This time out he would leave little to chance: After the experience with his insubordinate lieutenant, he would forego a second in command.

Halley was determined not to let the *Paramore*'s unresolved defects interfere with accomplishing his mission. He wholly believed that on his return his mission would "answer the expectations of those who perhaps censure the performance of my last voyage, without examining all the circumstances."

When he set sail in late September from the Downs with "a very

fine sky and a curious gale," it was well into hurricane season, but he
brimmed with confidence, the court-martial behind him. The
Falconbird, a Royal African Company ship, had escorted him past the
most treacherous sections of the African coast. On reaching Madeira,
the crew preferred to stay at sea to going to port. "People chose rather
to go without their wine than to lie beating in so much danger of the
Sally rovers," Halley wrote.

But fate tilted the odds once again. The next day, a vengeful squall
arose from nowhere, as if to play with the ship before it reached a
friendlier harbor. Suddenly, a large wave smacked the side of the
Paramore. Halley's cabin boy fell overboard. "We brought the ship
immediately a stays and heaved out an oar," Halley recorded in his
journal. Manly White was barely a teenager and a favorite of Halley's.
The crew kept searching for two hours, "but the sea being high and
the ship having fresh way, we lost sight of him and could not succor
him." A solemn Halley acknowledged the failure of his rescue attempt
in his journal. "At 6, I bore away south west, not thinking it advisable
to stay turning to windward in the latt of Madeira."

The loss of a man overboard was usually a devastating event for
all shipmates. As a later seaman once described: "There is a suddenness in the event, and a difficulty in realizing it, which give to it an air
of awful mystery. . . . The man is near you—at your side—you hear
his voice, and in an instant he is gone, and nothing but a vacancy
shows his loss. Then, too, at sea—to use a homely but expressive
phrase—you miss a man so much. . . . There is always an empty berth
in the forecastle, and one man wanting when the small night watch is
mustered. There is one less to take the wheel, and one less to lay out
with you upon the yard."

And Halley was probably more sensitive than the average sailor.
The drowning of his devoted cabin boy would scar Halley's psyche
for life. Though toughened by his time at sea, Halley "was so deeply
affected with the loss that during his whole life afterwards he never
mentioned it without tears," according to accounts from friends and
20th-century biographer Eugene MacPike. The tragedy was more

than Halley had bargained for when he undertook the second voyage. If the dreadful mid-October event distracted him from his scientific quest, necessity demanded he and the crew overcome it.

IN HALLEY'S DAY, DIVINATIONS OF THE heavens guided voyagers on Earth. Astronomy and navigation were inseparable. Sailing much beyond sight of landmarks required knowledge of the sky and the magnetic compass. To figure out where one was at sea (in terms of latitude), navigators would approximate a position by calculating the angle between the radius of the location and axes fixed in the Earth.

Location was also inexorably tied to time. The relative positions of the Sun, stars, and planets comprise the intricate inner workings of a celestial timepiece. But to tell accurate time using the mechanics of heavenly bodies at sea required clear visibility, a level deck, smooth seas, and at least several hours to calculate it. Plus, astronomy's clock was only partially understood theoretically speaking. And it worked only at certain times, depending on whether the navigator had the appropriate data at hand to make sense of it. For example, to use the Sun's altitude in a calculation, a mariner had to know how the polar distance to the Sun varied with the seasons. From the perspective of Earth, the Sun's path tilts in relation to the equator.

Halley was far better at making such calculations than most sea captains, who often relied on knowledge from previous trips, mere guesswork, and muddled musings. Even among scientists, Halley stood out as a meticulous observer. His credentials as an astronomer and his patient respect for scientific method placed him among the best navigators in the world.

Latitude, for the most part, was relatively easy to determine. And because many voyages traversed virtually due east to west or west to east, a captain merely had to find the right latitude and stay the course. Navigators of the time were versed in the procedure: Latitude was simply 90 degrees minus the angular distance to the North Pole. They also knew how to pinpoint latitude and local time from the altitude of the Sun, provided the distance to the North Pole was known.

Longitude, by contrast, befuddled mariners at the time of Halley's adventures at sea. But certain techniques worked under favorable conditions. Whenever possible along his sailing route, Halley calculated longitude at sea by employing several methods (all made easier when in sight of land). Most often he used dead reckoning with plots of latitude. He also relied on two astronomical clocks: determinations of lunar distances and timings of the eclipses of the satellites of Jupiter.

From the Moon's position in the sky, Halley obtained some guidance, although the Moon's irregular motion, which then was poorly understood, limited the accuracy of such determinations of longitude. But to succeed in his mission to chart the natural fluctuations in the Earth's magnetic field, accuracy was critical.

In his writings, Halley revealed how he made such observations. "From whence, at all times, under the like situation of the Sun and Moon, I might . . . have the effect of exact lunar tables capable to serve at sea, for finding the Longitude with the desired certainty." When other methods failed him, he used this method, as he describes: "Having by my own experience found the impracticability of all other methods proposed for that purpose, but that derived from a perfect knowledge of the Moon's motion."

Observing the eclipses and relative positions of the moons of Jupiter and the planet itself was often more helpful—at least when measured on the steadiness of landfall. Galileo had recognized this several decades earlier. On Halley's journeys, however, he used Jupiter's moons as clocks only when at port, not at sea, because of the difficulties viewing them from a tossing and pitching ship. But he told how he made such observations aboard ship when necessary:

> I had found it only needed a little practice to be able to manage a five or six foot telescope, capable of showing the appulses [near approach] or occultations [hiding of a bright object by passage of a body in front of it] of the fixed stars by the moon, on shipboard, in moderate weather; . . . whereas the eclipses of the satellites of Jupiter, how proper so ever for

geographical purposes, were absolutely unfit at sea, as requiring telescopes of a greater length than can well be directed in the rolling motion
of a ship in the ocean. . . . So in the remote voyages I have since taken to
ascertain the magnetic variations they have been of signal use to me, in
determining the longitude of my ship, as often as I could get sight of a
near transit of the Moon by a known fixed star: And thereby I have frequently corrected my journal from those errors which are unavoidable in
long sea reckonings.

Although Halley managed to observe from the deck of a moving
ship in mild conditions with a short telescope, using the longer scope
often met with trouble. On his first voyage in March 1699, when at-
tempting to observe a satellite of Jupiter from the coast of Brazil, for
example, he failed to catch sight of its eclipse because the "great height
of the planet, and want of a convenient support of my long telescope,
made it impracticable."

And then a month later, in Barbados, he encountered more difficulty. He was observing an immersion of Jupiter's first satellite when,
"the wind shaking my tube, I was willing to get a more covered place
to observe in, that I might be more certain, but when I again got sight
of the planet, the Satellite appeared no more."

Still his longitude calculations at various ports throughout much
of his journey were the most accurate to date. In fact, some of his
calculations, such as those at Bona Vista and St. Iago at 23 degrees, 0
minutes west longitude, and 22 degrees, 40 minutes west longitude,
respectively, reflect the modern true values. However, once he crossed
the equator at the African coast, his measurements were off by about
half a degree per day because he failed to factor in westerly equatorial
currents, which were imperceptible to him.

IT WAS THE VOLUBLE SEA—indifferent and, as Halley experienced from
the loss of his cabin boy, unforgiving—that shaped much of Halley's
scientific agenda. At the top of his wish list, Halley wanted a fast local
fix for navigation, available day or night no matter what the cloud
cover. Probably the biggest grail was longitude. By turning his atten-

tion from the heavens to Earth's mysterious magnetism, Halley had hoped for big things. No one had known what exactly to expect.

Halley had eyed the quivering needle within his 52-foot *Paramore* and recorded, as best he could, the reading from magnetic north. It was off kilter from true north. How much off kilter depended, of course, on the ship's location. On his first voyage, Halley made his first observation in Portland Road on the way out of the English Channel, recording the magnetic variation with great diligence. He made a single measure of the magnetic amplitude at sunset. He allowed for refraction on the horizon and then compared the result with the Sun's computed azimuth from geographical north.

What pattern might emerge from all these scattered readings at points in the North and South Atlantic? His first mission, though yielding promising results, had left the question open-ended. This time out his directive built on the scientific objectives of his first journey. Halley believed it might be possible to calculate longitude from the observed distance between magnetic north and true north. In this way, the magnetic variation could be used to identify those vexing but important lines oriented in a north-south fashion, whose spacing widened with the bulge of the earth.

What's more, this Royal Society fellow believed such a method could be universally true—good at all locations on the globe. It clearly worked for certain spots. And, best of all, the method would not require knowing the accurate time locally (an important point since even the best human-made clockworks kept poor time) or at the pole or reference point.

NAVIGATION WAS NO small matter. About the time of Halley's birth, England began operating under the economic idea of mercantilism. The principle held that for a maritime nation to increase its wealth the value of the country's imports could not exceed its exports. Colonization of overseas lands became the easiest way to offset negative trade balances with other countries. In 1650, 1651, and 1660, the English Parliament enacted the Navigation Acts, which by restricting

England's colonial trade exclusively to English and Irish ships helped establish the nation as a commercial power in Europe. The trade winds carried a plethora of tobacco, sugar, and other stores from her colonies in the Americas and West Indies. The supply of such commodities was steady enough to meet the needs not only of the upper classes but of the hoi polloi. Necessity had supplanted luxury. To finance the import of goods, countries began manufacturing goods for export. The flurry of trade would lay the foundation for the Industrial Revolution.

Of course, the Navigation Acts also served to fill the ranks of the Royal Navy at wartime and to protect trade routes. About the time of Halley's sea adventures, England's sea enterprise controlled some 4,300 ships weighing in at 340,000 tons and employed 55,000 seamen and at least another 50,000 hands to build and maintain the vessels. Though diminutive in comparison to today's tankers, most ships averaged 400 tons and were big in comparison to Halley's 80-ton pink. They typically averaged one round-trip voyage to the West Indies or North America each year. Only a dozen or so even larger ships would venture to the East Indies and China—generally a two-year endeavor.

The slave trade, in particular, raised the stakes for solving the longitude problem in the late 17th century. To maximize profits from the round-trip voyage from England to Africa to the West Indies and back to England, merchants exchanged slaves for sugar and other commodities. Such routes could not be easily achieved by merely sticking to a line of latitude.

Halley, arguably a seaman in his own right, had long been interested in advancing England's nautical prowess. In a letter written in 1695 to Samuel Pepys then retired as Secretary of the Admiralty, the administrative head of the Royal Navy, he criticized the "imperfect measure of knowledge in our ordinary navigation." For example, he noted that on a cloudy day sea captains could rely only on soundings to determine their position. In his letter Halley recognized the talents of such peers as Sir Anthony Dean, whom he considered "the great master of naval skill. No man can more scientifically give or account thereof than he." But Halley believed that navigation on the whole was still more guesswork than science. He disparaged "the absurd way of keeping their reckonings by the Plain Chart," as opposed to using the Mercator, which supposes the meridian is parallel and had been available for use on ocean passages for some 60 years. "In time of war, this erroneous manner of keeping the keel of the ships way by the Plain Chart occasions the loss of many ships," he continued.

International politics conspired to make his mission even more alluring. The Navigation Acts that enabled England to prosper also set off 60 years of feuding with Holland over supremacy of the sea. The first war erupted in 1652, roughly five years before Halley's birth. France, under Louis XIV, also entered the contest to control the world's oceans. The French first allied themselves with the Dutch to wage war against England in 1666. But the coalition fell apart two years later when Louis XIV tried to claim control of the Spanish monarchy by right of his marriage to a Spanish wife. His mounting aggression troubled the Dutch. In 1672, France and England joined forces to invade Holland.

Within two years England would withdraw from the conflict. The 1688 revolution of English Protestants against the Catholic King James II would lead to England and Holland forming an alliance. The bond was strong enough to squash France's bid to control ocean commerce. In this manner, the sea, or at minimum the human drive to dominant the oceans, would help change the course of English science.

ONCE AGAIN PAST MADEIRA, the *Paramore* and its crew pushed on. Eight days after the October accidental drowning, they made the island of Sal, one of the northern Cape Verde islands. Once ashore, Halley encountered a "Portuguese there with some few Black servants, who assured us there was no salt to be had, the salt pens being all in the water, we had leave to hunt and our people killed and brought aboard two Cabritos [an antelope variety], one very fat with good meat. We saw a turtle in the road, and turtle tracks on shore so four of our hands stayed on shore this night, and the next morning they brought off two turtles. . . . The Portuguese promised me some salt but made me stay till evening for it, and it was not much above a bushel when it came, bad salt mixed with dirt."

They sailed on from the island of Sal following the same course as the first voyage but absent the draining tension. They passed Bona Vista and then anchored at St. Iago. At both sites Halley successfully measured the longitude. From St. Iago, Halley reported back to London that all was well with the mission and that he was satisfied with his new officers. However, the native Portuguese were not all that hospitable. Halley bought the favor of a lieutenant and his sergeants for the "value of 5 dollars" and "none of my folks had anything taken from them," Halley noted in his journal. But ashore there was no market to be found. And those very same officers "would not let the botswain dispose of cheese or let us go to the country to buy fowls. This treatment I was obliged to put up with, on the accord of my small force, perhaps more men and guns might have procured better usage of them."

In early November "a smart tornado put us under our course," Halley recorded, and it was not long before the constellations changed to include the brilliant Southern Cross. On December 4, Halley managed to take a longitude reading because of a celestial event: the Moon passing in front of a fixed star in the Virgo constellation. "The moon did exactly touch this star with her southern limb at 3:15 in the morning at 3:20:20. The southern horn was just 2 minutes past the star. Having carefully examined this observation and compared with former observations made in England, I conclude I am in True Longitude from London at the time of the observations at 30:15."

Then on December 9 they "all smelled a very fragrant smell of flowers which the wind brought of the land; and several butterflies flew on board." They reached Rio de Janeiro on December 14 and entered the mountainous harbor then under the governorship of Artur de Sa e Meneses. The *Paramore* stayed in this enchanting port for two weeks as Halley again enjoyed Portuguese hospitality and the pleasant summer of the Southern Hemisphere. On the 29th the *Paramore* managed to leave the charms of Rio for the most dangerous part of the voyage: the search for *terra incognita*.

THE *PARAMORE'S* GEAR included a large brass sextant, quadrant, log and sounding lines, a pendulum clock (to tell time at night) that audibly tick-tocked, and two azimuth compasses. Above all, the magnetic steadiness of the compass was key in calculating longitude and determining if a search was being efficiently conducted. It was central to Halley's toolbox and hypothesis. And because a compass or two were standard equipment on all seagoing ships—no mariner would be without one—Halley knew his work could have immediate impact.

Familiar with Elizabethan physician William Gilbert's writings circa 1600 entitled *De Magnete* and the leading discussions since, Halley sought to better understand the invisible forces that made the needle swing. A contemporary of Shakespeare, Gilbert was the first to appreciate the reason a magnetic needle aligns itself roughly north or south along the meridians: terrestrial magnetism. He saw Earth itself as a gargantuan magnet.

Following on Gilbert and others, Halley elaborated on magnetism in a series of papers published in preparation for his journey aboard the *Paramore*. In a 1692 paper he hypothesized that Earth has four poles: two in its outer shell and two in its nucleus. In between a fluid medium might separate the inner and outer globes. He suggested that the outer shell rotated slightly faster than the inner one, accounting for the gradual shifting of Earth's magnetic field.

By the time of his second voyage, Halley had on board a barometer and a thermometer (the Fahrenheit scale was still about 20 years off). Finding the cause of atmospheric pressure changes was among Halley's many laboratory pursuits before his voyages. In his day the cause of the phenomenon remained a mystery. According to his logs, he recorded temperature and pressure daily with the latest devices. He carried one of Hooke's ingenious inventions: a marine barometer, specifically designed for use at sea. With it Hooke, indeed a bit overoptimistically, believed that "the observations of the weather may perhaps in great measure be timely enough discovered by the inquisitive and diligent mariner." Halley, however, backed Hooke's claims. He would later tell the Royal Society: "It never failed to prognostic [sic] and give early notice of all the bad weather we had."

In essence, the *Paramore* was the world's first floating observatory. Its cache of scientific instruments and books was superlative. Halley ported that long 20-foot refracting telescope and possibly even his 24-footer aboard the *Paramore*, scholars believe, even though such scopes were somewhat ungainly on land, let alone in a tossing, saltsprayed ship. But to Halley it was essential for viewing the moons of Earth and Jupiter. (Science historians, however, believe a 15- to 18-foot scope could have successfully captured the satellites. This minimum length was necessary to minimize spherical and chromatic irregularities.) According to his accounts to the Royal Society, Halley modified his telescope with a reflecting plate, an innovation first conceived by Newton to counter distortions inherent in such refracting scopes.

His portable arsenal of devices couldn't compete with the accuracy of the reigning permanent observatories of the day, especially those in Greenwich or Cassini's Paris. Many of Flamsteed's observations, for example, were remarkably precise even by today's standards.

Halley also took with him a library of sea tables and star catalogs. The books included records of the Sun's declination used to calculate latitude and tables of the Moon's position and Jupiter's position for finding longitude.

Luckily for Halley, London was England's center of navigational know-how. It housed most of the country's printers, publishers, authors, and cartographers as well as most if its makers of fine navigational and scientific instruments. Instructors versed in the art of navigation taught their science-based craft in the city on the Thames. Halley had grown up with access to the best nautical tools. About one-third of England's merchant ships—yet half the nation's tonnage—ran out of London.

For the most part, commercial sea companies were enlightened: Using the latest in navigation techniques was the best and least expensive insurance. And Halley desired similar coverage for his missions.

The books he brought with him probably included the newest edition of *The English Pilot*, issued in 1690. The ambitious book was the work of London's famed instrument maker and chart publisher, John Seller. First published in 1669, it included a sea atlas that detailed the world's seas and assorted volumes on specific seas. It was modestly subtitled *Sea Waggoner for the Whole World, with Charts and Draughts of Particular Places, and a Large Description of All the Roads, Harbors, and Havens, with the Dangers, Depths and Soundings in Most Parts of the World*. By all accounts, it was a vast improvement over the waggoners in circulation. These combined nautical manuals and chart books, first introduced by their namesake Dutch seaman Lucas Waghenaer in 1585, became almost obligatory at sea for all captains. The southern version included maps and directions for the Atlantic islands that Halley was passing in this leg of the mission: the Madeiras, the Canary Islands, Cape Verdes, and the Azores.

By the time Halley sailed, one of Seller's rivals, William Fisher, had come out with an *English Pilot* of his own that covered America and the West Indies. It was updated in 1698 and would have been an essential tool for Halley. By the time of Halley's voyages, the list of available titles was long. In 1685 Captain Daniel Newhouse published *The Whole Art of Navigation*, with a newer version in 1698; in 1686 James Atkinson published *The Seaman's New Epitome*; and in 1695 the mathematician Samuel Newton (no relation to Isaac) came out with *An Idea of Geography and Navigation*. Halley also likely carried a copy of the latest *Atlas Maritimus*, which depicted the South Atlantic, and a copy of William Dampier's *A New Voyage Round the World; Describing Particularly the . . . Islands in the West Indies, the Isles of Cape Verde, the Passage by Terra del Fuego . . . and Santa Helena*, which had just been published in February 1697, and many others.

Lieutenant Harrison's *Idea Longitudinis; Being, a Brief Definition of the Best Known Axioms for Finding the Longitude* had been available at booksellers since 1696. It's probably safe to assume there was not a copy of that on Halley's second voyage.

Of London's five chart specialty shops, Seller's was the most interesting and well known. The pioneering entrepreneur made and sold naval instruments and taught how to use them. He procured sandglasses (hourglasses) and sea compasses for the Navy Board. And along with fellow chart maker Joseph Moxon, he served as hydrographer to the king. Nearly a dozen other London shops also sold nautical maps.

IN LATE JANUARY 1700, THE THERMOMETER on the *Paramore* plummeted toward the freezing point. The shallow-bottomed vessel lurched southward to the edge of existing charts. Even the least-schooled sailor aboard had sensed danger was mounting as the thermometer continued to drop. Yet none could have suspected the true nature of the terror that lay ahead. Nearby, the ominous black fins of a pod of passing killer whales or perhaps bottle-nosed whales—both mammals until now the stuff of mere legend or perhaps an occasional incredulous sea monster tale—made fitting harbingers of the unknown threat.

As they hedged southward the weather became worse and worse. On January 18, Halley wrote in his journal: ". . . after midnight it blew extremely hard at N.W. with much lightning, and again a little before sunrise with very terrible lightning which seemed to be just on board us but God be thanked we received no damage by it." A cold current had replaced the balmy one that had pushed them down the South American coast. The weather continued to grow progressively more violent. And the temperature continued to drop. In the seclusion of Halley's cabin, ". . . which had been kept from the air, the thermometer stood but 11 above freezing."

The water appeared white, and small drifts of seaweed approached the ship. Assorted sea birds flew about. Halley and the crew reckoned that land was near. Then on January 27 a thick fog engulfed the *Paramore* and would hover for several days. They were likely south of the Falkland Islands. When the great fog dissipated, they saw "several fowls which I take to be penguins . . . being of two sorts; the one black head and back, with white neck and breast; the other larger and of the color and size of a young Cygnett, having a bill very remarkable hooking downwards, and crying like a bittern." Soon enough other creatures surfaced: "We have had several of the diving birds with necks like swans pass by us, and this morning a couple of animals which some supposed to be seals but are not so; they bent their tails into a sort of bow . . . and being disturbed showed very large fins as big as those of a large shark the head not much unlike a turtle." The fog was almost gone now. The cornucopia of flora and fauna befuddled even his most experienced officers. Halley pondered turning northward, but his course was set.

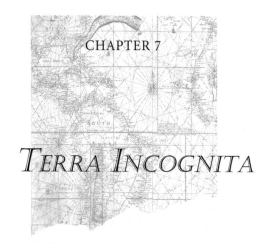

CHAPTER 7

TERRA INCOGNITA

The *Paramore* pushed toward the Antarctic, well beyond
Magellan's path through Tierra del Fuego. Outside the pro-
tection of this enchanting island cluster off the southernmost tip of South America, the cleavage of South Atlantic and South
Pacific often produced tricky winds and agitated currents. But the
ship passed without incident. The novelty of sea creatures and terrain unfolding before them probably only stoked the crew's wildest
imaginations.

On January 31 the *Paramore* reached the farthest point south ever
recorded by an expedition at the time, latitude 52 degrees, 24 minutes
south, but for the first time Halley distrusted his measurements. Another bed of weeds—often a sign that land or shallow waters were
near—floated by the ship. How unnerving it must have been for a
meticulous scientist like Halley. A commander does not generally like
more than the usual uncertainty.

On the horizon, an unusual, massive form emerged. Captain
Halley wrote in his journal, despite his stiff, cold hands, "flat on top

and covered with snow. Milk white with perpendicular cliffs all around." Three of them—monstrous, barren forms. None appeared on any of the charts in his collection. Did any lips dare to murmur it? Were they witnessing the first glimpse of mystical *terra incognita*? They seemed otherworldly. Halley held any speculations in check. "The great height of them made us conclude them land, but there was no appearance of any tree or green thing on them." But the crew quickly tried to introduce some familiarity. "Our people called [the first island] by the name of Beachy Head, which it resembled in form and colour. And the [second] island, in all respects was very [much] like the land of the North Foreland in Kent." Halley sketched the three masses and their positions in his journal, labeling the nebulous blobs A, B, and C. Undoubtedly by now, slick layers of ice had coated the *Paramore*'s masts and rigging, rendering the sheets hard to grip and to maneuver.

Halley estimated the height of the first island to be close to that of the real Beachy Head in Eastbourne Downland. At 530 feet it is England's highest chalk sea cliff. The second island was about 200 feet high, roughly the height of the real North Foreland lighthouse, which was established in 1499 and marks the southerly entrance to the River Thames. And according to Halley's journal, the second was "not less than five miles in front."

"The cliffs of it were full of blackish streaks," Halley described, "which seemed like a fleet of ships standing out to us. Wind blowing fresh, and night in hand, and because our vessel is very leewardly, I feared to engage with the land or ice that night, and having [stood] in as far as I durst, I resolved to stand off and on till day, when weather permitting I would send my boat to see what it was."

As if the crew was not already facing enough of a mystery, the night before them quickly turned foggy. Not until the next day at quarter past noon did the thick curtain lift. To Halley's eyes the landscape had rearranged itself. The islands were not as they had been at last sight. The first island now glared before them. "We saw the island we called Beachy Head very distinctly to be nothing else but one body of ice of an incredible height."

The voyagers had ventured farther south than any previous explorer in recorded history—to nearly 53 degrees latitude but to little avail. The strange masses proved not to be an ice-enshrouded *terra incognita*, but naturally occurring formations unfamiliar to experienced seafarers in 1700. By the time Halley realized the so-called islands were colossal floating icebergs, his *Paramore* was in harm's way. Once again, the world had proven not to be as it seemed—even the visible world of their common senses betrayed the crew and their elite navigator and diligent scientist.

"We were in imminent danger of losing our ship among the ice," Halley scribbled soon after in his journal, "for the fog was all the morning so thick that we could not see for long about us."

He immediately swung the *Paramore* northward, narrowly escaping, at least for the moment, wrecking in the fatally frigid water. He noted in his journal entry on February 1: "True course to this day noon is S 44 E 25 miles. Difference of longitude 29 minutes East; Longitude from London 35 degrees 13 minutes." Until that day the *Paramore* had never abandoned its course. The ship ventured within 13 degrees latitude of the Antarctic Circle and approached the northernmost tip of the Antarctic continent, just north of South Georgia Island, which was then undiscovered. They undoubtedly reached the Antarctic convergence, which surrounds the continent. In this region cold surface water moving away from Antarctica meets warmer, southerly moving surface water. Where the water masses collide, the colder water sinks below the warmer water. The net effect is a sudden drop in water temperature and salinity. Halley had just missed discovering a continent.

February 2. Fog set in again. Visibility was less than a mere furlong. Sometime after 11 a.m., the ghostly outline of another iceberg emerged on the *Paramore*'s leeward bow. Halley shifted his pink to avert collision. But now the *Paramore* found itself cruising on a path toward an even larger mass. Halley attempted to tack again, but the ship failed to move into the wind's eye. The poor draw of her square rigging disappointed her crew once again. Her captain knew full well that the icy mounds above the water were only an intimation of what loomed beneath the surface. Seven-eighths of each ice mass was submerged, an invisible threat to the most experienced seafarer.

"But the sea being smooth and the gale fresh got us clear: God be praised. This danger made my men reflect on the hazards we run, in being alone without a consort [that is, a companion or escort ship], and of the inevitable loss of us all, in case we staved our Shipp which might so easily happen amongst these mountains of ice in the fogs, which are so thick and frequent here."

Thanks to a sudden favorable gale that seemed to swirl in from nowhere, the explorers retreated unscathed to warmer waters. Halley felt he had done his best to satisfy the queen's desire to find *terra incognita*. Any further exploration might jeopardize both his crew and his primary mission.

Ultimately, Halley failed to claim Australia or Antarctica for the British monarchy. *Terra incognita* would remain elusive, although later explorers would know where not to venture at this time of year in the Southern Hemisphere. England would receive no bonus of doubling in size or of miraculous new elixirs from Halley's expedition. It would be nearly 70 years before another explorer would pass this way again to stare down the floating white beasts that guarded the passage to the seeming bottom of the globe.

Halley headed north toward his old stomping grounds of St. Helena, where he'd spent that seminal year as a student, and made for Tristan da Cunha, a remote, uninhabited cluster of islands of volcanic origin midway between South America and South Africa. The largest, nearly circular in shape, was reputed to harbor a towering fall of fresh water and exotic species of plants unique in all the world.

The fog, cloud cover, and chilly temperatures persisted. "The weather being commonly foggy with so penetrating a moisture that our linen, our clothes, our papers, etc. feel wet with it even in our cabins," he wrote in his journal. But at last February 8 brought "pretty clear sunshine."

After another week of sailing, an uncountable swarm of birds and albatrosses about the ship signaled the *Paramore*'s proximity to Tristan da Cunha. Its Portuguese discoverer had never set foot on the shore because he was unable to find a suitable landing place. Halley would not have any more luck.

They had been at sea for six weeks, the longest stretch of the voyage. Until now, Rio de Janeiro was the last land they'd glimpsed. Meanwhile, pitch-black clouds shrouded the main island's 1,800-foot peak, and thick fog and hazy air obfuscated the basaltic cliffs along its coastline. Often the dangers for seamen were just as great as at sea when a ship anchored close to shore. For one thing, maneuvering in a small boat in churning surf or along a rocky coastline could be perilous. Ever cautious, Halley decided not to attempt watering there and headed for the Cape of Good Hope.

While en route in the chill Benguela current, a violent squall erupted on February 26 that "threw us so that we had liked to have overset, but please God. She righted again."

Although leaks, perhaps aggravated by cold waters, had spoiled six dozen loaves of bread and 30 pounds of flour and cheese, Halley refused to return directly to England "fearing to go home in the winter, which would expose my weak ship's company to great hardships."

On March 11 they made St. Helena. From there Halley penned a letter to the secretary of the Admiralty, Josiah Burchett, informing him of their confrontation with the icebergs and his continued optimism for the mission: "In this whole course, I have found no reason to doubt of an exact conformity to the variations of the compass to a general Theory, which I am in great hopes to settle effectually."

The island was a refuge for Halley in several ways. Besides being a haven for some rest and relaxation after a long stretch at sea, it came to be considered "one of the most healthful places in the world," and sailors "when carried ashore here, recover to a miracle, rarely any dying though never so ill when brought ashore," according to an 18th-century ship's log. The East India Company had even made medical services available on St. Helena to passing seamen.

And here in 1677 Halley became the first person in the Southern Hemisphere to observe the transit of Mercury across the disc of the Sun. And here he likely developed the idea, which he first suggested in 1679, that the distance from Earth to the Sun (now known as an astronomical unit or AU) could potentially be calculated with a high degree of precision by observing the analogous transit of Venus across the Sun since it is the closest planet to Earth between it and the Sun. Because Earth and Venus orbit in different planes, this celestial alignment occurs only at intervals of 8, 121.5, 8, and 105.5 years. Not only could humankind have a better idea of the scale of our solar system, but also with such knowledge could improve star charts for navigation at sea.

In a 1691 paper Halley presented his idea to the Royal Society on using transit timings from several distant locations on Earth to accurately calculate the Earth to Sun distance. Readings from different sites would help astronomers pinpoint the timing of the end and the start of the egress, the period when Venus moves inside the limb of the Sun. By observing the transit at different locations that are known distances apart, the Earth to Sun distance could then be determined by triangulation. Because the Sun appears to have a diameter 30 times that of Venus, the silhouette of Venus seen from different vantage points on Earth appears against different points on the Sun due to parallax. The greater the number of simultaneous observations, the more accurate the calculation would be.

More than 60 years before Halley's presentation, Johannes Kepler had determined relative distances between known planets. For example, scientists knew that the distance from the Sun to Mars was slightly more than 1.5 times the distance between the Sun and Earth. But absolute distances between celestial bodies were still unknown. Kepler estimated the distance between Earth and the Sun to be 24 million miles, which was off by a factor of almost four.

Halley reasoned that the differences in the apparent paths to Venus's transit could be used to determine the distance between Earth and Venus. From this information, basic trigonometry would solve the distance between Earth and the Sun.

Actually, Scottish mathematician James Gregory had suggested the idea in passing several years before Halley. But it was Halley who advanced the idea and promoted sea expeditions to observe the transit of Venus across the visage of the Sun from distant sites around the world. A pair of English astronomers, Jeremiah Horrocks and William Crabtree, made the first known observations of a Venus transit in 1639, predicted by Kepler. It appeared as a black spot roughly one-thirtieth the diameter of the Sun. Though visible to the naked eye, they viewed it indirectly using small telescopes that projected the image onto paper. But the Englishmen lived only 30 miles from each other, Horrocks in Lancashire and Crabtree in Salford. Many more measurements would be needed to determine the distance from Earth to the Sun. But another transit wasn't due to occur until 1761.

Because of the rarity of the event (at least measured in the human life span), the transit wouldn't be sufficiently observed until 1769, almost a century after Halley's writings and decades after his death. A Royal Society commission secured King George III's approval for an expedition to Tahiti to do exactly that. Based in large part on Halley's proposals, the mission was led by Lieutenant James Cook aboard the *Endeavour*. According to Cook's journal, the day that he measured the transit from the French Polynesian island "proved as favorable to our purpose as we could wish, not a cloud was to be seen all day and the air was perfectly clear."

Cook's transit data in combination with observations from 63 other locations around the globe enabled astronomers to calculate the distance from Earth to the Sun more precisely than ever before. They carefully deduced a distance of 95 million miles. The figure came within a few percentage points of the value accepted today: about 93 million miles. (The number wasn't officially agreed upon internationally until 1942.)

Cook also continued southward where Halley left off and made the east coast of Australia—that southern land of intrigue, nearly 70 years after Halley almost met his demise amid the icebergs. More than a century would pass before Antarctica would be discovered.

The goliath intellects of Copernicus, Kepler, Galileo, and Descartes chiseled the physical science foundation of Halley's day. Halley's accomplishments as an astronomer were shaped not only by the resulting contemporary world view but also by the instruments available to him.

At the time of Halley's birth in about 1656, the Copernican revolution in astronomy was at least a century old and nearing acceptance in mainstream thought. About this time at the French Academie, the Dutch vanguard Christiaan Huygens built the first efficient pendulum clock, key for measuring time at night at remote locations. By the time Halley began studying astronomy, it was well established theoretically that the Sun—not Earth—was the center of our solar system. By then, viewing sights had also been introduced to telescopes, but not without controversy. Although the old observing guard clung to their established methods using open sights, telescopic sights would soon prove superior. They afforded a 40-fold enhancement in precision over the instruments used, for example, by Tycho Brahe, the most accurate observer of the 16th century. The first telescopes, made in the Netherlands, didn't come into use until about 1609. Galileo Galilei in Venice was among the first astronomers to point one at the heavens. Halley himself would be active in improving the observing technologies of his era in several ways. Nonetheless, the technology on hand in Halley's day was good enough that he was able to advance stellar astronomy.

Although questions about the size of the universe and the population of its stars intrigued Halley, his discovery of stellar motion was probably his most important accomplishment in that field. Always attune to history, Halley decided to take a look at Ptolemy's observations from the 2nd century A.D. Since before Ptolemy's time, stars had been looked on as stationary with respect to each other, though they moved in a circular fashion around the North Star. But when Halley scrutinized Ptolemy's star catalog, he noticed some irregularities. Even after compensating for errors, Halley realized that the flagrant discrepancies between Ptolemy's charts and ones put together some 1,500 years later meant the stars must have moved in relation to each other or independently. He was able to determine the individual or "proper motion" of three very bright stars: Arcturus, Procyon, and Sirius. Halley surmised that the motion of brighter stars would be easier to detect because they were closer to Earth. Conversely, the farther away the star, the harder it would be to detect its proper motion. The technology to expand on his premise wouldn't be developed for another 150 years.

The limitations of the precision of the instruments then available also prevented the successful measurement of a single accurate stellar distance in Halley's lifetime, though that didn't stop Cassini and others from making triumphant claims that they had made accurate determinations of such distances.

UNTIL CONSTRUCTION OF THE GREAT observatories in London and Paris in the 1670s made English and French astronomers famous, Johannes Hevelius, a city councillor and Consul of Danzig, was the most renowned astronomer in all of Europe. Besides being among the first to observe transits of Mercury and Venus, he discovered several comets and the first known variable star. He mapped the Moon's surface and proposed that comets follow parabolic orbits.

The son of a prosperous brewer, Hevelius built an observatory on his family's property that was probably the grandest and best equipped of any at the time. Curiously, he made most of his observations from the roof of his townhouse. He corresponded with other leading astronomers, including Flamsteed the astronomer royal and Cassini in Paris, and was made a fellow of the Royal Society.

When a dispute erupted in the 1670s between Robert Hooke and the then-elderly Hevelius over the question of open sights versus telescopic ones, the Royal Society called on Halley to settle it. With his recent experience at St. Helena and his penchant for diplomacy between some of the high-strung egos in science, he was an ideal choice.

Rightly proud of his achievements and reputation as an observer, Hevelius refused to adopt the emerging advances in instrumentation. He preferred to use his proven sights: pinnules or metal plates with small holes in them that were attached to his quadrant. Stubbornly, he dismissed out of hand the improvements in precision afforded by telescopic sights, which had yet to yield comparable results.

At first glance, Halley might hardly be considered impartial. After all, telescopic sights were behind the success of his original catalog of 300 stars completed on his St. Helena trip during his Oxford student days. In fact, Halley was among the first to use the new methods in the field. (Of course, Flamsteed and Cassini from their perches at the world's leading observatories had little incentive to survey the heavens from the field at all.)

In late May of 1679, fresh from his early triumph at St. Helena, Halley visited the like-minded Hevelius at his invitation. They began observing on the first night of his arrival in the magnificent Baltic city. Hevelius would describe Halley as "a very pleasant guest, a most honest and sincere lover of truth."

Although Halley stayed for two months, he was satisfied—nay, dumbfounded—with the accuracy achievable with Hevelius's instruments. He reported his findings to the Royal Society within the first 10 days of his visit: "As to the exactness of the . . . distances measured by the sextants, I assure you I was surprised to see so near an agreement in them, and had I not seen, I could scarce have credited the relation of any; verily I have seen the same distance repeated several times without any fallacy agree to 10 seconds," or 1/360 of a degree.

In fact, on some observations Halley and Hevelius, working to find star positions in tandem on a six-foot sextant with both a movable and a fixed sight, attained an accuracy within five seconds, or twice as accurate as Flamsteed claimed with his telescopic sights. In Halley's absence, Hevelius often observed the night sky with his second wife, Catherine Elisabeth, a reputed beauty 36 years his junior.

In this period, women were rarely engaged in science. For a start, even the brightest women were denied access to any sort of formal or disciplined education. But there were some exceptions. Flamsteed's wife Margaret Cooke, whom he married later in his career in 1692, also assisted him with compiling astronomical data and mathematical calculations.

The use of two observers likely increased the open-sighted sextant's accuracy, as did Hevelius's customization of the graduation of his instruments. The Danzig astronomer and Halley also observed with telescopic sights that Halley had brought. The best they achieved was an accuracy within 10 seconds. Hevelius believed this added additional support to his argument, but the shortfall was likely due to the fact that Halley's instruments, being portable, couldn't be expected to generate the same accuracy as the best larger scopes permanently stationed at a big observatory, which remained still and in calibration. Hevelius's largest scope was 150 feet long. While important for power, length, of course, did not necessarily govern accuracy. Flatter-profiled lenses could potentially achieve better clarity.

Halley was impressed enough that with prompting from Hevelius he wrote a testimonial. "I offer myself voluntarily as a witness of the scarcely credible certainty of your instruments, against anyone who shall hereafter call the truth of your observations into question," Halley wrote. Hevelius's observations may have been more accurate at that time and place, but telescopic sights would soon be shown to unequivocally afford finer resolution and greater accuracy than the unassisted human eye.

But Halley's kind letters did little to persuade Hevelius's persistent adversaries: Flamsteed, Hooke, and others. The controversy thundered on, forcing Hevelius on the defensive. Hevelius charged that Halley's visit was more about espionage than diplomacy. He was dispatched "for no other purpose but to rigidly examine [my] instruments."

Back in London, the usually reserved Halley apparently took offense at that charge and responded less moderately: "I am very unwilling to let my indignation loose upon him . . . for I would not hasten his departure by exposing him and his observations as I could do and truly as I think he deserves I should."

Then a devastating tragedy would leave the squabble forever unresolved in Hevelius's mind. A mere two months after Halley left the hospitality of his host, Hevelius's observatory went up in flames, destroying most of his instruments and observations along with the facility. The fire was allegedly started by a careless servant whom Hevelius later called "the most perverse animal on two legs."

News arrived in London that the fire had claimed Hevelius's life. In a 1681 letter to Catherine Elisabeth, Halley expressed his sorrow and offered to make good on a request to buy her a fashionable dress hand-tailored in London from 10 yards of the highest-quality silk. "We cherish ardent hopes that Mr. Hevelius may still survive and that rumour has shown herself false in this particular, though it is but rarely that she deceives us in unhappy things. I quite realize that his heartbroken spouse must be wearing sad-coloured apparel, yet for several reasons I have thought well to send the gown procured for her," Halley wrote. "Anyhow she will be able to preserve it until her period of mourning is past."

All of London learned several months later that Hevelius had survived the fire, but the catastrophe did not stop proponents of telescopic sights from raging on with their technology crusade, nor did it stop Flamsteed, when it would suit his political agenda, from later using the silk dress as evidence of some sort of indiscretion on Halley's part. His insinuations seemed plausible enough. After all, Hevelius's wife was only 10 years Halley's senior. Hevelius might have been her grandfather.

Although Halley remained in Hevelius's good graces thanks in part to his work at St. Helena, Hevelius continued his personal attack on Hooke: "That he makes it his own business to persuade him and all the world, that his own way is the best, safest, and most exquisite, which ever can be invented by any; reproaching this author all along for not obeying him and following his dictates (as if this author were one under his command) bragging only of what he can do, but doth nothing," Hevelius ranted in an issue of the *Philosophical Transactions*. Hooke's rivals in the Royal Society, including John Wallis, relished whatever collateral damage might result from such invective. Meanwhile, Halley cleverly let the proverbial chips fall as they may. And there was political mileage to be gained.

The Royal Society secretaries, Francis Aston and Tancred Robinson, who allowed Hevelius's harangue to be published, were forced to resign. Installed in their place within the next couple of months were John Hoskins and Thomas Gale, two friends of both Hooke and Halley. In turn, a salaried post of clerk was set up to which Halley was easily elected. Oversight of the publication of *Philosophical Transactions* fell on Halley's plate. While Hooke's reputation suffered somewhat in the process, Halley's prominence rose.

The sight controversy ended with Hevelius's natural demise in 1687. In death Halley still considered him a friend. A new era in observation was dawning, and Halley was in the thick of it. For Hooke it was just another instance where his contributions were key but where others who were better political navigators would reap greater rewards. In his diary Hooke expressed envy of Halley and his contemporaries' daring travels. Six months before Halley sailed to St. Helena, Charles Boucher had voyaged to Jamaica, enduring raging seas and shipwreck. Boucher lost his precious instruments and books, but he brought back hot new data from his observations afforded by the good seeing on the island. In 1687, Sloane would also travel to Jamaica, where he'd be introduced to cocoa, to serve for a time as physician to its new governor, the second Duke of Albermarle. And soon enough Halley, the risk taker, would again be "a sayling," as Hooke described it, and Hooke once more was left ashore.

AFTER THREE WEEKS OF REST on St. Helena, Halley set a course for his *Paramore* back across the Atlantic on March 30, 1700. The drinking water they had obtained on St. Helena was rather brackish because it rained heavily during their stay. So he set his sights on the island of Trinidad, roughly 900 miles off the Brazilian coast.

On the sail, Halley recorded latitude daily and measured the variation almost every day. He used his telescopic sights—no offense intended to Hevelius—to observe the Moon on April 11. He calculated the relative position in latitude between St. Helena and Trinidad to be 21 degrees 20 seconds.

Sure enough, three days later they glimpsed the rocky outline of the three isles of Martin Vaz and on the 15th reached Trinidada, which Halley called both Troindada and Trinidad in his log. "The island being about 1.5 leagues in length, lying nearest NW and SE. The North and West part is nothing but steep rocks scarce accessible."

Pleasantly surprised with the unpopulated piece of real estate once ashore, he thought the island, with its bountiful water supply, would be a valuable resupply point for English ships. Halley rowed around the island in the pinnace, and he and his crew took all of five days to map it. He measured the variation to be 6 degrees 30 seconds east. He planted a Union flag with its hallmark crosses and claimed the small oasis for the king of England (though the Portuguese discovered it in 1502). And the crew released a breeding stock of goats, hogs, and guinea fowl on the island.

The tiny island was hardly *terra incognita*. But for the Crown, it would have to suffice for now.

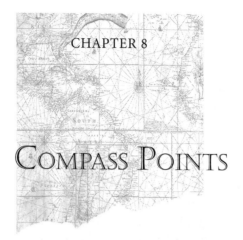

CHAPTER 8

COMPASS POINTS

alley's log April 29: Seven months into the voyage, the
Paramore reached the Brazilian city of Pernambuco
(modern-day Recife). Palm trees rustled along the coastal
port's white sand beaches.

The Portuguese governor, Fernao Martins Mascarenhas de
Lencastre, welcomed the ship's company to the provincial capital. He
happily informed the visitors that Europe was still at peace. To Halley
the assurance was "most desirous" news. They wasted no time pro-
curing wine and other staples.

Pernambuco was protected by a natural reef and distinguished
by versatile inland waterways inspired by those of Venice. The city
was in the process of becoming the busiest port in Brazil after the
Dutch departed in 1654, sparking an era of strong nativism.

The next day, when they tried to load the supplies aboard the
Paramore, they were stopped by a certain man named Mr. Hardwick.
Claiming to be the English consul, he accused the visiting sailors of
piracy. The ungainly rigging of Halley's pink had once again raised
suspicions. Hardwick had two of Halley's crew frisked, and he re-

quested Halley's commission and royal instructions. Those documents seemed to satisfy him; he let the interlopers continue packing the commodities. He showed an interest in Halley's unusual mission or at least feigned one and invited the captain to call on him at his house the following afternoon.

When Halley reached his door, Hardwick's men surprised him and arrested him. Hardwick still suspected Halley was a pirate. But after searching the *Paramore* stem to stern and finding nothing even vaguely resembling pirate loot or the weaponry of buccaneers, he released Halley and apologized. Hardwick claimed to be acting on behalf of the Portuguese. It was a charge the governor later denied despite the fact that he had furnished the guard during Halley's brief captivity.

From Fernao Martins, the governor, Halley soon learned that Hardwick was only posing. He was no consul and had no authority to jail Halley. It turned out Hardwick was actually a frustrated Royal African Company agent, whose own ship, *Hannibal*, had recently been hijacked by pirates in the vicinity.

To be sure, virtual lawlessness didn't exist only at sea. A 1665 account of law and order on Barbados, which was Halley's next port, is telling. The powers that be accused a successful but supposedly unbalanced planter named John Allin of blasphemy, which earned him a stern reprimand from the newly appointed governor, Lord Francis Willoughby of Parham. Taking offense, Allin swung at the governor with his sword, slicing off two fingers and gashing his forehead. Allin then committed suicide before reaching prison. To send a message, Willoughby had him drawn and quartered. His remains were "dry-barbecued, or dry-roasted, after the Indian manner, his head to be stuck on a pole at Parham, and his quarters to be put up at the most eminent place of the colony."

HALLEY'S INTERESTS WERE broad, even among 17th-century thinkers. No subject seemed too foreign for his scrutiny. In 1691, he published a paper on the place and date of Julius Caesar's first invasion of England on the basis of evidence of a lunar eclipse: From Caesar's description of a full Moon and the winds and tides on a particular night, Halley determined that the landing occurred on August 26, 55 B.C., at Deal. A 1695 paper touched on Palmyra, an ancient city in Syria. He posited that the inscriptions on an old Roman altar found there indicated that during periods in the past the Moon's motion had accelerated.

And the list goes on and on. In 1686 he published a paper on trade winds and monsoons in which he identified a key factor in their origin: solar heating. Though he acknowledged his explanation was incomplete, the paper would prove to be the first significant one in a new field, a field he would forever be credited with founding: geophysics. A map, which he added to the paper two years later, represented the first meteorological chart and headed a new direction in map making: concentration on a specific physical topic.

A few years later, in June 1690, Halley spoke before the Royal Society on the phenomenon of hurricanes. Such mundane events as thunder and lightning would not be explained for nearly a century. Besides observing that hurricanes occur in the latitude of the Caribbean Islands as well as other places like the China Sea and the Bay of Bengal, he may have been the first to point out the existence of a hurricane season. In another effort to link terrestrial events to astronomical conditions, he noted that hurricanes occur when the Sun returns "over the zenith of these places" in August or September.

Halley's curiosity about the heavens was rivaled only by his inquisitiveness about more earthly matters. His whole magnetic quest was just one dimension of his inquiries. Halley questioned all things related to the seas. Often it was experience that informed or at least augmented his mental musings.

In April 1691, Halley was called in to help salvage a frigate owned by the Royal African Company. Called the *Guynie*, she was fresh from the Gold Coast, carrying "184 elephants' teeth"—most likely ivory tusks—according to a bill of lading, and a large amount of gold. Valuable cargo at any rate. She had gone down near Pagham off the seem-

ingly calm Sussex coast. By chance, Halley happened to be in the area
experimenting for the Royal Navy on, of all things, diving bells.

Halley first wrote about diving bells in a March 1689 paper he
presented to the Royal Society. He knew of West Indian pearl divers
who could plunge to great depths to recover treasures by holding their
breath for seemingly impossible periods. He saw great possibilities
for salvage work if a better method might be devised "to walk on the
bottom at a considerable depth of water and to be there at liberty to
act or manage one's self to the best advantage as if one trod upon the
dry ground." And he set about to invent one: "Whereby a man might
have his bell as a house over his head, and stand on the bottom almost dry." To test his ideas he borrowed a frigate from the Admiralty.
Halley made his first dive in the bell off a tiny seaport on the south
coast near Sussex.

The hollow wood and copper cone required nearly two tons of
counterweight to submerge it. Once again Halley took great personal
risk for science. "When we let down the engine into the sea, we all of
us found at first a forcible and painful pressure on our ears which
grew worse and worse till something in the ear gave way to the air to
enter, which gave present ease," Halley wrote of his plunge. A 40-gal-
lon cask was lowered from the frigate to replenish air in the bell every
15 feet of the descent. The diving bell could be lowered to 10 fathoms
and enabled the divers to work comfortably inside at that depth for
up to two hours.

Halley formed a company to develop his invention, according to
a newspaper account that mentioned the promise of a diving stock:
"If Mr. Halley should succeed, of which (were the wars at an end and
the seas secure) he seems very sure . . . it would be very considerable."

Halley kept working on the *Guynie* salvage project until 1696,
when he was called away by Newton to work at the mint. Halley was
to help him correct a lingering problem with the nation's currency. A
common practice had evolved to clip the edges off silver coins and
melt down the shavings. The agreed-upon remedy was to replace the
old coins with newer ones with milled edges to prevent the clipping

and pilfering of silver. Halley spent three years in Chester as deputy controller. There he continued his scientific writing and observations, patiently waiting to embark on his debut journey aboard the *Paramore*.

During this interlude, likely much to his chagrin, King William had assigned Halley's new vessel to a visiting Russian czar named Peter, who was interested in learning about the latest in shipbuilding and mingling with great British minds of the day. Although the *Paramore* was built and designated for Halley's voyages, the Admiralty approved its use "as the Tsar should desire."

The emperor of Russia would later be known as Peter the Great. His seamanship, however, proved far from great. His so-called sailing experiments took their toll on the *Paramore*. Whether Halley actually sailed with the czar remains an open question.

But Halley apparently made the best of the needed refitting, conversing with Peter at length about developments of the day. He even accepted Peter's invitation to dine at his table and sample a fine brandy or two and did so again later, according to some versions of the story. Martin Folkes, a personal acquaintance of Halley and future president of the Royal Society, gave this account of the czar's visit: "Halley . . . possessed all the qualifications necessary to please princes who were desirous of instruction, great extent of knowledge, and a constant presence of mind; his answers were ready, and at the same time pertinent, judicious, polite and sincere."

On visiting England, the emperor of Russia, according to Folkes,

> sent for Mr. Halley, and found him equal to the great character he had heard of him. He asked him many questions concerning the fleet, which he intended to build, the sciences and arts, which he wished to introduce into his dominions, and a thousand other subjects which his unbounded curiosity suggested. He was so well satisfied with Mr. Halley's answers, and so pleased with his conversation, that he admitted him familiarly to his table, and ranked him among the number of his friends, a term which we may venture to use with respect to a prince of his character; a prince truly great, in making no distinctions of men but that of their merit.

Halley's choices in experimentation also revealed that he was as much a practical man as an intellectual. In particular, his dalliance with all things nautical or related to navigation is illustrative and helped build his sea credentials with institutions such as the Royal Navy as well.

In tandem with the diving bell, he designed a diving suit fitted with pipes to inhale and exhale air and a lantern that could be attached to the bell. He also concocted a primitive depth gauge and tested ways to protect the eardrum when descending and ascending to different pressures.

Halley had claimed priority for his invention:

> A Man having a suite of Leather fitted to his body, with a cap of Maintenance such as I have formerly described, capable to hold 5 or 6 gallons, must be perfectly enclosed so that the water may as little as possible soak in upon him, must have a pipe coming from the diving bell to his cap, to bring him Air, which must be returned by another pipe, which must go from the cap of maintenance, to a small receptacle of air placed above the Diving bell into which it is to return the Air, that has been breathed; whilest the other brings it to the man.

Always a profound thinker, Halley also experimented during this period with how sound and light behaved underwater. Newton used Halley's observations in writing his *Opticks*, which he published in 1704, only after his favorite nemesis, Hooke, had died:

> Of this kind is an experiment lately related to me by Mr. Halley, who in diving deep into the Sea in a diving Vessel, found in a clear Sunshine Day, that when he was sunk many Fathoms deep into the Water the upper part of his Hand on which the Sun shone directly through the Water and through a small Glass Window in the Vessel appeared of a red Colour, like that of a Damask Rose, and the Water below and the under part of his Hand illuminated by light reflected from the Water below looked green.

For thence it may be gathered, that the Sea-Water reflects back the violet and blue-making Rays most easily, and lets the red-making Rays pass most freely and copiously to great Depths. For thereby the Sun's direct light at all great Depths, by reason of the predominating red-making Rays, must appear red; and the greater the Depth is, the fuller and [more intense] must that red be.

And at such Depths as the violet-making rays scarce penetrate unto, the blue-making, the green-making, and the yellow-making Rays, being reflected from below more copiously than the red-making ones, must compound a green.

THE *PARAMORE* REACHED the British colony of Barbados on May 20, 1700, at about 5 o'clock in the afternoon. Here Halley found his Majesty's ship the *Speedwell* heading to sea for England. At full sail the 94-foot-long "fireship," which carried explosives and an armament of about eight small guns, must have been a familiar and somewhat soothing sight. In battle such a vessel was sailed as close as possible or even attached to a much larger and more heavily armed enemy ship and then ignited by a slow match and a train of powder. The captain and crew would then escape as fast as humanly possible in a smaller boat towed alongside the craft or astern.

The governor of Barbados, the Honourable Ralph Grey, Esquire, received Halley but advised him to leave immediately. The island was experiencing its most severe typhoid outbreak of record. Halley only permitted the number of men ashore required to replenish the water cask. On many port calls earlier in the voyage, Halley had successfully kept the crew aboard the ship to minimize the risk of his people contracting tropical or exotic diseases, which were often fatal.

But the governor's warning came too late. Halley and many crew members were stricken by the sickness. Its cause was then unknown, but in fact it was a bacterial infection caused by consuming contaminated food or water. "I found myself seized with the Barbados disease, which in a little time made me so weak, I was forced to take [to] my cabin," Halley managed to report in his journal.

From his ample berth, Halley ordered his first mate to set a course for St. Christopher's (modern St. Kitts), a four-day, 400-mile jaunt from Barbados. They got under sail in such a rush that they left the stream anchor behind in Barbados. It was used to secure the stern when the *Paramore* didn't have enough room to swing from the bower, or bow anchor.

In a later letter to Secretary of the Admiralty Burchett, Halley described the Barbados plague as "a severe pestilential disease, which scarce spares anyone and had it been as mortal as common would in a great measure have depeopled the island."

Disease actually was a greater peril at sea than wrecking or work-related accidents. Each year for every thousand English sailors who shipped out, roughly 5 would suffer an accidental death, and 10 would drown in shipwrecks. But more than 45 would succumb to disease.

The leading killers were dysentery and malaria, but there were a multitude of other deadly diseases out there, which varied from region to region. In the Caribbean alone the list included yellow fever, dropsy, leprosy, yaws, and hookworm.

Halley was aware that the crew's health was everything. Given the fiscal realities of the Royal Navy, Halley must have pulled some strings to get a surgeon aboard his undersized *Paramore*. He likely knew about the most pressing health threats, from books like William Cockburn's 1696 *Account of the Nature, Causes, Symptoms . . . Distempers . . . Incident to Seafaring People.* Perhaps due to his influences, Halley's surviving son, also named Edmond, who was born while he was captaining the *Paramore,* grew up to become a Royal Navy surgeon.

Of course, sea surgeons of the day could do little more than sea captains in the face of most diseases. Though they had some success treating ailments like syphilis with mercury, healing wounds and injuries was their forte. Bloodletting was much the era's rage in medicine, even though it usually made the sick sicker. And sailors—even the most educated of captains—would often treat themselves. As William Dampier, the adventurer and scientific observer, who sailed at the time of Halley's mission, recalled in his journal, "I found my fever to increase, and my head so distempered that I could scarcely stand. I whetted and sharpened my penknife in order to let my self blood; but I could not, for my knife was too blunt."

Unknown to Halley, his contemporary Dampier had succeeded in finding a large landmass in the South Seas, reaching Australia one year before Halley's first mission (and some 70 years before Captain Cook). He sailed aboard the pirate ship *Cygnet*. He would return two years later in 1699 aboard HMS *Roebuck*, reaching Dirk Hartog Island in western Australia. The *Roebuck* would wreck on its return voyage to England, however, sinking off Ascension Island in 1701. Although Dampier's relationship with the Admiralty would be marred, he and his specimens would survive.

Dampier and Halley had many similar interests and were likely acquainted through their connections with Samuel Pepys. Though more of a self-schooled maverick, Dampier was fascinated with nature, navigation, and magnetism. On his subsequent Admiralty-backed voyage to explore the Australian and New Guinea coasts, he also compiled observations of magnetic declination in some areas not mapped by Halley. He published his readings in *A Voyage to New Holland*.

In this period, progress in medical science greatly lagged that in physics or astronomy. People of all classes and educational backgrounds adhered to superstitions, magic, spiritual rituals, and the like. Even someone as erudite as Pepys believed in amulets, which some people still use today to ward off disease. He credited his "fresh hare's foot" worn around his neck with dispensing a spate of good health.

Sailors often were versed in sundry remedies learned on their travels. For example, Jesuit's bark, or cinchona, was known to help thwart malaria. Later researchers would learn it was a source of quinine. Cannabis and opium were easily purchased over the counter in London or at port. Opium was touted as a "panacea." Marijuana, however, came with a recommendation: "The seed, which heats and dries, and by much using abates seed in man, cures coughs, asthma, jaundice and other like diseases," according to a popular medical book available to commoners.

Little was known of personal hygiene and sanitation. Food and water were readily contaminated by a host of bacteria. The transmission mechanism of most diseases remained a mystery. Since fleas and lice happily inhabited the wardrobes and wigs of London's wealthiest denizens, it's a given that they plagued sailors, too, as flea-infested rats scurried aboard a great many ships.

The cure for scurvy was not known at the time of Halley's mission, though fortunately none of Halley's crew was reported to have contracted it. However, they typically were at sea no longer than six weeks, the amount of time required for its onset. For less fortunate sailors, after about three months of subsisting on standard Navy rations, they would develop full-blown symptoms, including fatigue, depression, bleeding gums, reopening of wounds, and painful, incapacitating swelling of the joints.

Available statistics reveal that mortality at about the time of Halley's voyages was in general significantly higher on Dutch ships than on English ones. Edward Barlow, a sea captain who traveled most sea routes between 1659 and 1703 and even had a run-in with Captain Kidd and his *Adventure Galley* in May 1697, speculated that this was because "English ships commonly make shorter passages and are better provided with provisions." Ironically, this was the only instance in his *Journal* that he had anything remotely positive to say about the food served on English ships.

Ordinary rations included salted beef and pork, dried cod, cheese, butter, and dried peas. Aboard the *Paramore*, the ship's cook made biscuits from flour or bought bread at port. Watery beer was the everyday beverage, but they bought wine and rum at assorted depots along the way when possible. Many sailors also had their own stash of brandy tucked away in their sea chest. They supplemented their diet with fresh fish, fowl, and other wildlife caught or hunted off the ship or purchased ashore, including turtles and perhaps an occasional monkey.

Portrait of Edmond Halley in a Royal Navy captain's uniform was painted by Sir Godfrey Kneller when Halley was presumably in his early 40s, about the time he set sail aboard the *Paramore*. Source: National Maritime Museum, Greenwich.

Portrait of Sir Isaac Newton painted by Sir James Thornhill in 1712. The unusual portrait, which shows Newton without his wig, is one of fourteen completed before his death in 1727 by various artists. Source: Courtesy of the Trustees of the Portsmouth Estate. Image by photographer Jeremy Whitaker.

Portrait of Halley painted by Thomas Murray around 1687. Halley was about 30 years old at the time and was the clerk of the Royal Society. Source: The Royal Society.

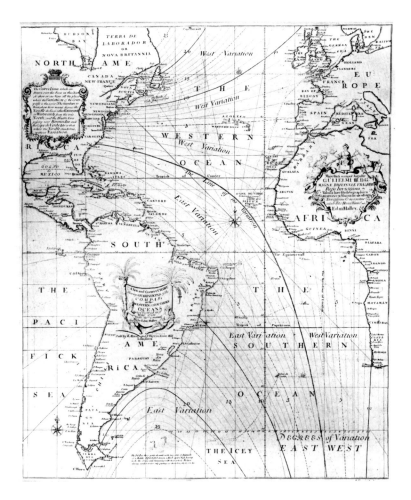

A 1701 version of Halley's Atlantic sea chart depicting lines of equal magnetic variation of the compass. Halley added a written description to later editions of the chart that is reprinted in the Appendix. Source: Royal Astronomical Society.

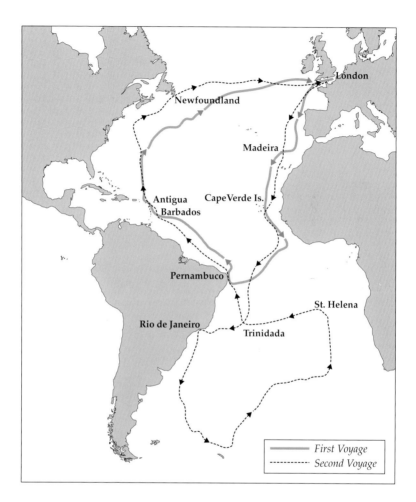

The routes of Halley's first and second voyages on the Atlantic aboard the *Paramore* in 1698 and 1699-1700, respectively. Pernambuco is modern day Recife, and Trinidada is modern Trinidade, Brazil.

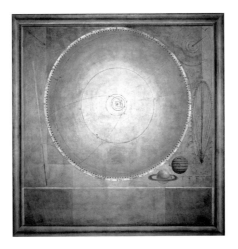

A fresco depicting the orbit of Halley's Comet conveys a 1770 understanding of the solar system. Halley's namesake reappeared in 1758 to the awe of the world but 16 years after his own death. Source: Meridian Room at the Museum La Specola in Padua, Italy.

An engraving of Halley's diving bell and helmet, one of a few surviving illustrations. Although he published a description of his invention in a 1689 edition of the Royal Society's *Philosophical Transactions*, this image appeared in a more popular publication decades after his death. His underwater adventures are described in Chapter 8. Source: W. Hooper's *Rational Recreations*, 1782, from the British Library collections.

Danzig astronomer Johannes Hevelius and his wife Catherine Elisabeth observing with a sextant. The two-person instrument was likely the same one that Hevelius also used with Halley when he visited in 1679 as detailed in Chapter 7. Source: Hevelius' 1673 *Machina Coelestis* at the Royal Astronomical Society.

Photographs of the Royal Observatory, Greenwich, today. The core buildings, completed in 1675-1676, remain much as they were when Charles II built the facility for the first Astronomer Royal John Flamsteed. It is the site of the Prime Meridian and also where Halley's capstone was relocated in 1845. Source: National Maritime Museum, Greenwich.

Halley was probably more aware of proper dietary needs than the run-of-the-mill captain and most likely acquired fresh fruits and vegetables at port as well. It is not known whether he was aware of the emerging wisdom that lemon, lime, or other citrus juice seemed to ward off scurvy. Lemons were packed aboard many Royal Navy ships at this time, logs reveal, and English sailors were eventually dubbed "limeys."

PERHAPS HIS WILLINGNESS to tempt death inspired Halley to develop the first mortality tables in 1692, though he had dabbled now and again in population problems. Or perhaps it was the high number of children he and his wife, Mary Tooke, lost in infancy. Halley had "changed his condition," on returning to London after spending the whole of 1681 in Italy. He and Mary are said to have lived together "very happily and in great contentment" despite such losses, which were almost expected given the staggering infant mortality rate in the late 17th century.

Whatever his inspiration, when the Royal Society received the bills of mortality for Breslau, Germany, from Henri Justel, whom he had first met in 1681 in Paris, Halley took an interest. (Breslau was then the capital of the province of Silesia.) These bills listed the age and sex of all Breslau citizens who died between 1687 and 1691. For example, of the 6,193 infants born, about 1,740 died before age 1 and 3,460 survived to age 6. Although similar data were available for London and Dublin, Halley believed they were skewed by fluxes in immigration that interfered with recordkeeping and increased adult mortality rates as evidenced by "the great excess of the funerals above the births."

Halley devised a table giving values of annuities for up to age 70 for the presumably more stable Breslau population. He used five-year intervals (which are still used by contemporary insurance companies). He thought the time period adequate, "leaving it to the ordinary arithmetician to complete the calculation, whenever bills of mortality should be given for a suitable large number of years." He assumed the sample was representative of the larger population, a premise that is key not only for producing actuarial tables but for other population studies too.

Halley's interest was deeper than mere mathematics, nonetheless. He was captivated by human eventuality. In commentary accompanying the table, he wrote: "How unjustly we repine at the shortness of our lives, and think ourselves wronged if we attain not old age. Whereas it appears hereby, that the one-half of those that are born, are dead in seventeen years time. So that instead of murmuring at what we call untimely death, we ought with patience and unconcern to submit to that dissolution, which is the necessary condition of our perishable materials, and of our nice and frail structure and composition: and to account it a blessing, that we have survived perhaps many years that period of life, where at the one-half of the race of mankind does not arrive."

He also noted that if more women were married, the population could grow more quickly; for example, four out of every six women could have a child each instead of one of every six:

> The political consequences hereof I shall not insist upon; but the strength and glory of a King consisting in the multitude of his subjects, I shall only hint, that above all things celibacy should be discouraged, as by extraordinary taxing and military service, and those who have numerous families of children to be countenance and encouraged by such laws, as the *jus trium liberorum* [a law granting privileges for families with three or more children] among the Romans; but especially by an effectual care to provide for the subsistence of the poor, by finding them employments whereby they may earn their bread without being changeable to the public.

His observations are somewhat telling. Along with showing his awareness of history, intrigue with the study of social relationships, and flair for analysis, they reveal how some elements of his class limited his views. Though broad-minded as a scientist, he apparently viewed poverty as a societal burden more than a social injustice.

Halley's ideas about estimating life span for annuities and other purposes—thus placing a value on an individual life—were generally met with contempt, regardless of their political interpretation in terms of population management and unemployment. The first complete work on annuities, by a French mathematician who lived in England for most of his life, Abraham De Moivre, wouldn't be published for almost 50 years. Halley was recognized posthumously for the priority of his contribution when the Amicable Society's charters were published in 1790, some 85 years after its founding. In fact, for the most part, all future actuaries adopted Halley's general formula for computing annuities and use of tables to display the results of calculations.

WHILE DEATH REMAINED a mysterious phenomenon in Halley's lifetime, marriage was a straightforward proposition: It was more about transferring property than romance. The more prosperous a family, the more money mattered in the transaction, which may explain why it wasn't difficult for him to spend months away at sea from his family to pursue his ambitions, at great personal danger. No letters from Halley to his wife during his voyages survive, but she was likely on his mind from time to time. During dinner and what is called "Saturday night at sea," in the vernacular, British naval officers toasted their "Sweethearts and Wives." In July 1675, Henry Teonge wrote of the tradition in a Saturday entry of his experiences aboard the HMS *Assistance*: "We end the day and week with drinking to our wives in punch bowls."

Whether Halley married purely for love is doubtful. It appears that he returned to London from his tour of Italy at age 25 for the nuptials, which likely were arranged by his and the Tooke families and their lawyers, which was customary among the upper classes at the time. Wooing was minimal in such arranged marriages.

Although Halley had already achieved status for his celestial charts and came from a prominent family, in social terms he may have bettered himself. The Tookes were at least the Halleys' social equals if not their slight superiors. On his death in 1663, Mary's father, Christopher, willed her and her sisters money and lands, as did her Uncle Edward Tooke, who passed on in 1668. Her grandfather James was auditor of the Court of Ward and Liveries. Her father, uncles, and grandfather were all Inner Temple lawyers, members of a high-powered society of barristers headquartered in the heart of London, which made for a very litigious family.

Halley married his bride in the church of St. James at Duke Place in 1682. Prominent citizens were married there in private ceremonies under a special license granted by the government. St. James was one of three churches that claimed exemption from the jurisdiction of the Church of England, as a so-called peculiar. Every day up to 12 such private marriages were performed at St. James. Halley received 200 pounds, her share of her family legacy, for marrying Mary.

Under the canons of 1604, most marriages were conducted in public at one of the party's parish churches. Banns, giving notice of pending vows, were read on three consecutive Sundays. More common couples were married in their parish churches, usually the bride's. Of course, some of the well-to-do relished the spectacle of a showy public wedding.

But increasingly, rich and poor alike resented the invasion of privacy afforded by the banns. In general, the wealthy preferred to keep themselves above public comment, given the financial terms of their contracts. And for the lower classes, often the bride or groom's youthful indiscretions were public knowledge, and the couple was subjected to raucous laughter from the congregation. In those days premarital sex or "bundling," that is, heavy petting, was commonplace and part of many courtships.

Clandestine marriages, replete with sketchy paperwork, were also available to the lower classes, however, in dozens of small marriage houses off the maze of alleys at the base of Ludgate Hill near Fleet Bridge. The cost was equivalent to a week's wages for the average working man. The backdrop for these seedy wedding chapels was fittingly the Fleet Prison and the Fleet Ditch, a mucky channel brimming with rotting sewage and fish parts that fed into the Thames. The area came to be called the Rules of the Fleet.

The practice became increasingly popular among Londoners at the time of Halley's voyage despite a law that was passed in 1696 that fined clergy 100 pounds for marrying a couple illegally without proper banns or licenses. At the turn of the century, roughly a third of all marriages were performed near the Fleet. The ceremonies were valid in the eyes of the church provided some conditions were met. The practice, however, fostered abuse, enabling close relatives to marry and almost encouraging heiresses to be abducted and forcibly wed and raped.

Daniel Defoe castigated "the arts and tricks made use of to trepan and as it were kidnap young women away into the hands of brutes and sharpers [which are very scandalous], and it become almost dangerous for any one to leave a fortune to the disposal of the person that was to enjoy it and when it was so left, the young lady went always in danger of her life." The practice wouldn't be outlawed until 1753 by the Marriage Act, more than a decade after Halley's death.

Halley probably thought of his family often during his bedridden days battling typhoid. But the often-fatal infection did not win. After a week's respite on St. Christopher's in the Caribbean, good fortune found Halley once again. All members of his crew seemingly miraculously recovered "by the extraordinary cure of my doctor," Halley wrote. Although Halley survived, the illness, which sometimes causes a rosy-spotted rash on the chest and abdomen, blemished his fair skin. "Though it used me gently, and I was soon up again . . . it cost me my skin."

AND THE MISSION pressed on. From St. Christopher's, it was Anguilla and then back to Bermuda. The contagion past, Halley ordered the *Paramore* refitted for the return voyage. "Our decks and upperworks being leaking by being so long in the heats, I hired three caulkers to assist my carpenter, who in six days had finished their work, and I gave a new coat of paint to our carved work, which was very bare and parched." After the overhaul, Halley acquired a new mate named George Tucker. His original right-hand man, Edward Sinclair, had jumped ship so to speak, taking a better position with a commercial captain.

Nothing could keep Halley from his observations. Not gales, personnel matters, icebergs, or stray gunfire. Almost every day he observed latitude and variation, noting his course and the winds. Longitude he necessarily calculated more infrequently.

Halley now guided his *Paramore* to the coast of North America. He charted a direct course from Bermuda to New England, specifically, the great hook of Cape Cod. After reaching the island of Nantucket, conditions were such that Halley opted to make for Newfoundland.

The passage took 10 days, and they arrived through heavy fog. "Had we not fell in with some French fisherman, we might have been on shore," Halley remarked in his journal.

But danger still lurked. Up the Newfoundland coast, the *Paramore* made the Isle of Sphere (modern Cape Spear), which Halley's men dubbed the "Isle of Despair." Approaching nearby Toad's Cove, they encountered a small armada of English fishermen. At first sight of the *Paramore*, the fleet headed downwind—a common evasive maneuver. As Halley pointed his ship into the harbor, a series of shots flew across the bow, grazing some of the shrouds and ratlines but failing to damage the masts or sails.

Once again the odd rigging of Halley's pink had been taken by jittery sailors to be that of a pirate vessel. His men apprehended the gunman, a fisherman named Humphry Bryant from Bideford, Devon, who reported a pirate vessel had recently plundered one of his fleet. Forgivingly, Halley let Bryant go.

The *Paramore* would soon be homeward bound. Halley and his crew could afford to be magnanimous.

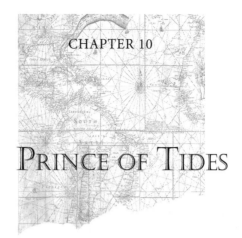

CHAPTER 10

PRINCE OF TIDES

Before the ink on his chart had barely dried, the Admiralty dispatched Halley on yet another mission. This one was not just for the sake of science. It involved the main artery of the English realm.

In the spring of 1701, Halley set out once again in the *Paramore*. He was to chart the tides and other vagaries of the English Channel, and in the furtive role of spy he was also to survey assets near French ports on the south side of the channel, which the French call "La Manche."

Once again Halley had drawn up his own orders and leveraged the monarchy to support another scientific voyage:

You are to use all possible diligence in observing the Course of the Tides in the Channel of England as well as in the mid sea as on both shores, and to inform yourself of the precise times of High and Low Water; of the set and strength of the Flood and Ebb and how many feet it flows in as many places as may suffice to describe the whole. And where there are irregular or Half Tides to be more than ordinarily curious in observing them. You

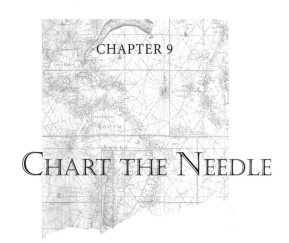

CHAPTER 9

CHART THE NEEDLE

The *Paramore* arrived home on September 10, 1700, to Deptford, almost a year to the day after its last departure. If anyone had expected royal fireworks, they would have been disappointed. There weren't even any special notices in the London newspapers. Halley's final entry in his ship's log was perfunctory: "We delivered our guns and gunners stores and the pilot being on board by low water we weighed from Long Reach and delivered the pink this evening into the hands of Captain William Wright, Master of Attendance at Deptford."

Halley soon arrived in London. The odd bounty he carried, gleaned from nearly two years and many thousands of undulating miles, was a clutch of numbers. Never had anyone shown as much interest in the wide blue sea—not merely the shoreline—and brought back little else but abstract data. These nearly 150 observations ranged from roughly 50 degrees north latitude to 50 degrees or so south latitude. He hoped they would be translatable into something worthy of the expense. If the readings seemed less immediately impressive than a haul of gold or spices, Halley needed more time and hunched over his data to prove otherwise. Even to the thinkers at the Royal Society, it was not yet evident what Halley had accomplished.

By February 1701, after months of work, Halley issued his best results. As many ways as there were to stare at the numbers and coordinates, no new overarching theory of magnetism emerged. He would not be able to publish a tidy equation for magnetism as Newton had done for gravity. But in connecting the dots, the plots of similar declination values, patterns had emerged. Halley had converted his daily observations into curved lines depicting areas of equal magnetic variation. Looking a bit like an alien spider web, the chart showed for the first time the invisible magnetic field that swaddled Earth—or at least some of its features.

If hardly earth-shattering to those scientists who were hoping for a new theory, Halley's results were immediately useful to sailors and to the large practical world of trade. (And the findings did nothing to upset his earlier theory.) He proudly and boldly made his curved lines as prominent as those of latitude and longitude. Although the execution may have appeared confusing at first, the chart superimposed the invisible (yet real) magnetic lines over the artificial common grid of mariners. And in this one step he boosted the utility of the grid, increasing the probability of finding one's way, at any point along the way, within this global latticework.

In a sense, if a map shows you how to find your way, this was a "metamap," one that reduced your chances of getting lost. Halley probably anticipated that most sailors would find the chart intimidating. The usually modest man felt the need to call out its novelties. "What is here properly new, is the Curved-Lines drawn over the several seas, to show the degrees of the variation of the magnetical needle, or sea compass," Halley wrote in a description in the top corner of his chart (see Appendix). "They're designed according to what I myself found in the Western and Southern oceans in a voyage I purposely made at the public charge in year of our Lord 1700."

Halley called them "curved lines." Soon they came to be called

"Halleyan lines." To this day they remain the leading method for de-lineating this information (though we now call them isogonic lines, a more encompassing term that would come into use a century after Halley's mission).

His 1701 magnetic sea chart debuted to immediate success. The course of Halley's journey aboard the *Paramore* was etched into a single 22- × 20-inch sheet. It was the first printed and published map to employ isolines to indicate a specific value of a particular measurement, in this case magnetic declination. John Harris did the engraving. His chart went on sale at the Postern on Tower Hill by a recently launched partnership of Richard Mount and Thomas Page.

This was the culmination of his career and involved a wealth of skills. And Halley played it as far as he could. The chart also traced Halley's voyage into iceberg-infested waters, visits to exotic territories, and stops at St. Helena, Rio de Janeiro, Bermuda, and Newfoundland. Pictures hinted at his encounters with new species. Halley made no mention of his escapades with Lieutenant Harrison, however.

Halley was so proud of his enhancement of the Mercator representations that he thought it should be given a special name. "The Projection of this Chart is what is commonly called Mercator's; but from its particular Use in Navigation, ought rather to be nam'd the Nautical; as being the only True and Sufficient Chart for the Sea." But in this case, his proposed name didn't catch on enough to displace the 16th-century innovation of Flemish geographer Gerardus Mercator.

Time would prove that Mercator's projection, which purposefully distorts the pattern of latitude and longitude lines to always cross at right angles, would in fact best represent magnetic field declination on navigational charts. On such maps the distances between parallels of latitude generally increase from the equator toward the poles proportionately with spacing of longitude meridians. Rhumb lines, or lines of constant compass bearing that ships typically steer along appear as straight lines. The major disadvantage of this type of projection is that polar regions appear especially enlarged.

This combination of new English isolines overlaid on the old
Flemish projection became popular across Western Europe. His chart
was commonly pasted in editions of the *English Pilot* and became a
best-seller. There was no Battle of the Books regarding its utility.
People recognized the fact that it made compass navigation more accurate and the world a little less mysterious. It was almost as if all sea
pilots, formerly nearsighted, suddenly viewed the world in crisper
detail. Paths at sea from now on would be more efficient and approaches to shorelines safer.

Halley's achievements, the charts, were praised in his lifetime. On
Halley's return, Dr. John Wallis, Savilian Professor of Geometry at
Oxford and a long-time member of the Royal Society, wrote that his

> magnetical chart fixes the business of the magnetic variation, in these
> seas for the present time. If similar observations had been made in former
> age, and transmitted to us, it would have been of great use. And if such be
> made in future, from time to time, and recorded; by which it may appear
> at what rate the variation varies; it will afford great insight into the mag-
> netic doctrine, about which we are now so much in the dark. . . .

> I believe that Dr. Halley has more than any other applied the deepest
> thought in establishing a theory of the declination of the magnet which
> the learned have been seeking to improve ever since; yet he does not ven-
> ture to determine by geometry the situation of the magnetic poles upon
> the earth and to establish rules for computing the declination. Mean-
> while however, he has empirically constructed crooked lines representing
> the declinations of the magnet on the largest ocean of the world, and he
> has had the good fortune to see those lines, which were constructed
> mostly on the basis of the observations taken during his voyages con-
> firmed more and more by later experiments.

Halley dedicated that first chart, which covered the Atlantic
(mistakenly judged the world's largest ocean by Wallis), to King William III.

The king never remarried after Queen Mary's death in 1694. Although during her lifetime he had an ongoing affair with one of her

ladies-in-waiting, Elizabeth Villiers, it was more of a ploy to win Mary's devotion. Despite the rough start of their arranged marriage, Mary, who was 12 years his junior, came to love him and his native Holland. His lasting grief after her death, however, revealed his true feelings of respect and admiration for her and explains his dedication to fulfilling her ambitions for England.

Fortunately for Halley, William rewarded him with a handsome bonus for his contribution just before his death in March 1702. The 200-pound authorization was equivalent to 12 years of wages for the average able-bodied sailor.

Since Halley's day and the dawn of global exploration, charting the changing Earth's surface field has been essential to world powers. Even almost two centuries later, for instance, Halley's charts were touted by G. Hellmann, a prominent German historian of science, as "a masterpiece of practical navigation." His representative system was embraced as being "of the greatest importance in all branches of physical geography."

Contour-type lines of equal field increments are still plotted today. And the science of determining one's continual geolocation within a grid has continued to make astounding leaps of accuracy—from high-tech military targeting to car navigation.

As William Gilbert had recognized at the time of Shakespeare, Earth itself acted as a giant magnet. Now, a century later, Halley had become part of that history. By trekking much of the aqueous globe, he had created a unique treasure map to share with anyone who wanted it.

GILBERT DIDN'T PUBLISH his textbook on geomagnetism until 1600. Ostensibly, among the first science books aimed at the public, the text attempted to separate myth from fact:

> In follies and fables do philosophers of the vulgar sort take delight; and
> with such like do they cram readers a-hungered for things obtuse, and
> every ignorant gaper for nonsense. But when the nature of the lodestone

shall have been by our labours and experiments tested, then will the hidden and recondite but real causes of this great effect be brought forward,
proven, demonstrated . . . and the foundations of a grand magnetic science being laid will appear anew, so that high intellect may no more be
deluded by vain opinions.

Iron's magnetic properties and its relative abundance on the
planet had lent credence to Gilbert's original idea that Earth was a
permanent magnet. Yet one of Gilbert's more important contributions was the notion of spherical influence, which would in turn spur
Kepler to develop his laws of planetary motion and eventually describe a magnetic field. Kepler was among the first to consider that
Earth's gravity and the Sun's gravity might somehow be "magnetic,"
though his ideas were somewhat wrongheaded.

The first Royal Astronomer Flamsteed, too, suggested magnetism
might produce gravity, the nature of which would remain very much
a mystery for some time. But his explanation of how was also off
base. Newton only scratched the surface of geomagnetism in his
Principia. He gave it little thought and simply noted it was likely "very
small and unknown."

In the second half of the 17th century, Royal Society pioneer Robert Boyle advocated the idea that magnetism was caused by a corpo-
real, atomic effluvium that was in constant movement. (Effluvium
was a pet theoretical device.) His improvement of the air pump
helped confirm that magnetic action occurs in a vacuum. Boyle argued that direct contact with an effluvium caused magnetic attraction—an idea that was compatible with Gilbert's permanent magnet
hypothesis.

A general belief emerged among English thinkers that variation
in magnetic properties occurred because of changes in the iron's tex-
ture due to mechanical operations. The concept supported Boyle's
notion of effluvium. To pursue this line of inquiry, the Royal Society
created a "Magnetics Committee," which compiled declination measurements and conducted various lodestone experiments. Annual

measurements of variation were made in London by other investigators as well.

Years before Halley's mission, faith in magnetic philosophy as an independent discipline was replaced by the notion that magnetism would be explained under the rubric of effluvial dynamics, essentially the movement of an invisible vapor comprised of the smallest but still tangible component of an element. This conclusion relegated magnetism back into the broader category of study—that of what would later be known as terrestrialism. Halley and others turned their attention toward deciphering the inner workings of the planet.

Despite decades of supposition, a leading question remained: how to explain the slow but sure motion of the lines of magnetic declination by a couple minutes or so of arc each year. Descartes, for one, supposed that iron ore deposits on the Earth's surface might account for the motion. But Halley had debunked such a notion during his 1676 trip to St. Helena. Other ideas, such as Robert Hooke's that the magnetic poles were moving in extremely slow circles several degrees off kilter from the geographic poles, also didn't hold up.

In his 1683 paper Halley suggested that "the whole Globe of the Earth is one great Magnet, having 4 Magnetical Poles or points of Attraction . . . and that in those parts of the world which lie near adjacent to any of those magnetical poles, the needle is governed thereby; the nearest pole being always predominant over the more remote."

His idea was that there were two north magnetic poles and two south magnetic poles. In a follow-up paper in 1692, he surmised that one of each must be on the Earth's surface. And the second pair must be located on an inner sphere, perhaps 500 miles beneath the surface.

When he presented his first paper on geomagnetism to the Royal Society, Halley ran an experiment for the society using a needle and two lodestones. He demonstrated that a compass needle follows the closest and strongest magnetic source. He divided the Earth into four imaginary regions, each ruled by a distinct magnetic pole. He proposed that the fourth was significantly stronger than the others, which added a layer of complexity. The poles were placed such that they were unbalanced enough that two separate antipodal dipoles would not result.

The setup would account for the variations in declination that he and others observed. If the inner and outer spheres rotate at different speeds, he reasoned, the changes would occur. Halley thought so highly of his theory that his portrait, painted at age 80, had a diagram of his spherical shells in the foreground.

In his day, Halley's initial paper on geomagnetism drew the ire of his one-time mentor Flamsteed. He believed Halley had borrowed key ideas from a little-known mathematician named Peter Perkins, whom Halley consulted on his deathbed in 1680. Perkins was master of a prestigious school for boys, the Royal Mathematical School at Christ's Hospital, established to produce a small cadre of highly trained navigators. Originally a hospital for friars, Christ's Hospital was converted to a charity school in the 16th century, and by the time of Perkin's arrival it had become London's largest school.

Flamsteed alleged: "His discourse in the former transactions concerning the variation of the needle and the 4 poles it respects I'm more than suspicious was got from Mr. Perkins. . . . He was very busy upon it when he died. . . . Mr. Halley was frequently with him and had wrought himself into an intimacy with Mr. Perkins before his death, and never discussed any thing of his 4 poles till sometime after I found it published in the transactions."

Halley never admitted any wrongdoing, let alone did he bother to defend himself. He let Flamsteed stew over his alleged "art of filching from other people and making his works their own." The dispute marked the beginning of the decline in their relationship. No one of renown substantiated Flamsteed's charges. It is known that Perkins was working on his own theory of variation, but there's no evidence it was similar to the one Halley presented.

At minimum, their clash over Perkins's data reveals an apparent difference in their worldviews. Flamsteed had a sense of proprietorship when it came to his research and demanded that his data only be disclosed at his discretion. Scholars attribute his attitude in part to his penchant for perfectionism. He didn't want to publish data prematurely that might still be enhanced. Prone to taking on large issues, Halley believed in essence in the value of sharing data and that the data of public servants, even if underpaid, belonged in the public domain. He actively sought out the results of others to develop hypotheses about not only geophysical but also historical patterns. Halley's method typically entailed compiling observations from varied sources and then publishing them openly in the *Philosophical Transactions.*

Flamsteed was relentless in his campaign against Halley. In a November 1686 letter to Richard Towneley, a colleague of Boyle's and fellow researcher known for his work on air pressure and safekeeping of Gascoigne's micrometer, he opined:

> He is got into Mr. Hooke's acquaintance, has been his intimate long and from him has learnt these and some other disingenuous tricks. For which I am not a little concerned for he has certainly a clear head, is a good geometrician and, if he did but love labor as well as he covets applause, if he were but as ingenuous as he is skillful, no man could think any praises too great for him. I used him for some years as my friend and I make no stranger of him still, but I know not how to excuse these faults even in my friends. Since he ran into Mr. Hooke's designs and society, I have forborne all intimacy with him.

Hooke, incidentally, had long grown tired of Flamsteed as well. After all, a decade before Flamsteed had branded him "a conceited cockscomb." Hooke had returned such salvos in kind, considering Flamsteed "proud and conceited of nothing" and an "ignorant impudent ass." And in more recent diary entries, Hooke—not once but twice—flat out wrote: "Flamsteed mad." Whether he thought Flamsteed insanely angry or a bit nuts remains open to interpretation.

In many areas of emerging science the fringe is often not far from the facts. Astronomers acknowledge their debts to astrologers, chemists to alchemists. Recall that even Newton pursued alchemy, and he continued to do so even after writing the *Principia*. Boyle was likewise fascinated with alchemy and corresponded with Newton on the subject. Over the centuries, the charms of magnetism attracted charlatans, as well as bona fide scientists like Halley, with remarkable consistency. Magnetics have been linked to everything from water witching (using a divining rod to locate underground water sources or metal deposits) to health remedies to the cause of shipwrecks in the Bermuda Triangle.

Petros Peregrinus, a French scholar and experimenter, wrote the first detailed descriptions of both floating and pivoted compasses in his 1269 *Epistolia de Magnete*, which was also the first western account. Peregrinus was convinced that a lodestone could be manufactured that would serve as an impeccable timepiece if it could just be properly cut and positioned. Shortly thereafter others proposed that a mechanical clock could be constructed to keep perfect time at sea and obsolete the astrolobe, a celestial planisphere that is rotated over a plate that divides the heavens into a network of bearings, or azimuths, and altitudes, or almacantars, according to the vantage point of an observer at a given altitude.

In the early 16th century, some, like Spain's Gemma Frisius of Louvain, believed clock technology had advanced enough to be carried at sea. When the notion reached England, the self-taught seaman William Borough tested the reigning watch technology. But such watches lost as much as 15 minutes a day and required seemingly constant winding. After less than a month at sea, that margin of clock error could translate to an error in longitude as large as an ocean. A successful timepiece needed to be accurate within seconds for several weeks at sea.

Robert Hooke was among the first to recognize that a pendulum-based chronometer would not work at sea. While Huygens was perfecting his pendulum clock at the French Academie, Hooke was toying around at Oxford with spring-driven marine chronometers that didn't rely on pendulums to set a clock's pace. Huygens, however, beat him to a patent on a similar spring balance clock in France in 1675. Hooke was somehow insulted while trying to patent his device in England. Henry Oldenburg, still a Royal Society secretary, cast one of the barbs, backing Huygens's prototype over Hooke's. Hooke abandoned the patent, precluding even a preliminary trial. Oddly enough, Hooke didn't appreciate the importance of the capability of determining longitude at sea. "No kind or state would pay a farthing for it," he declared.

While an eternal clock that worked at sea would for a time prove quite valuable to society, the concept of a perpetual motion machine powered by magnetism, first proposed by the very same Petros Peregrinus, would become a perpetual lark in the science world. In the 18th century, one true believer presented his ideas for a perpetual motion machine not once but twice to the Royal Society. He argued that it would run best in Barbados at the magnetic equator. There the intense horizontal component of the field would continually flip a bar magnet. The U.S. Patent and Trade Office continued to grant patents for perpetual motion machines well into the 1970s. To date, none has been built, but a few enthusiasts still hold out hope.

Peregrinus also proposed a magnetic levitation device. Jonathan Swift lost no time parodying supporters of the idea in *Gulliver's Travels*, first published in 1726. Once again taking a shot at the natural philosophers, Swift described a flying magnetic island he called Laputa that hovered over fictitious lands. He suggested that magnetic repulsion could act as "anti-gravity," suspending the island in space. Swift also wrote of Laputa's astronomers who calculated the periods of 90-odd comets. Again he was taking a satirical swipe at Halley, the return of whose comet 32 years later would naturally quell such ridicule.

A more personal use of magnetism developed shortly after Halley's death, when Franz Anton Mesmer, a Viennese physician, introduced magnetic remedies to the field of medicine. He claimed he could manipulate the magnetic fields that he believed ran through all living creatures. Although he enjoyed famous clients like Mozart, when Mesmer moved his practice to Paris, Louis XVI set up a royal commission in 1784 to look into this so-called animal magnetism. Written by the likes of Antoine Lavoisier, Benjamin Franklin, and Joseph Guillotin, the commission's report found that "animal magnetism may exist without being useful, but it cannot be useful if it does not exist." A series of controlled experiments revealed that Mesmer's art was nothing more than the power of suggestion. Nonetheless, the word "mesmerize" was added into the vernacular of many western countries, and his ideas would influence the development of hypnosis and chiropractics in the United States.

A late-18th-century magnetic quack, Dr. James Graham, encouraged upper-class newlyweds in London to spend their wedding night in a suite filled with tons of magnets with the promise of conceiving "super children." After his initial success, Graham's "Royal Patagonian Magnetic Bed" was found to be fraudulent.

Claims, which have continued to resurface over the course of recent centuries, that magnets offer health benefits have lost ground under modern scrutiny. Today, hucksters still advertise them to provide pain relief, but such remedies have yet to pass clinical muster.

Perhaps because he was just too busy, Halley, unlike his friend Newton, resisted prolonged departures into abstract mysteries. But Halley's theoretical quests often yielded practical ends.

ALTHOUGH HIS THEORY of geomagnetism never panned out as he intended, his chart offered a navigational tool for sea captains. Such was the demand—avant-garde mariners wanted to correct their compasses—that Halley began to expand the chart to cover more of the world. To make it more legible, Halley created a better explanation of the chart to accompany it.

In a marvelous display of scientific sharing of data, Halley, using data supplied by many others, expanded his famous chart in 1702 into Halley's World Chart of Magnetic Variations, "A New and Correct Sea Chart of the Whole World showing the Variations of the Compass as they were found in the Year MDCC." Despite its all-encompassing title, though, it wasn't entirely global—not enough reliable information existed to cover the Pacific—not to mention that on it California appeared as an island. The expanded chart was dedicated to Queen Anne's consort: "Prince George of Denmark, Lord High Admiral of England, Generalissimo of all Her Majesty's Forces."

Halley couldn't resist a parting gibe in the Battle of the Books saga. His new patron was Queen Anne, sister of his original patron, Queen Mary. In a Latin ode to Her Majesty on his world map, he contrasted the esteemed empires of Greek and Roman antiquity to that of modern Britain. Its Latin verses, translated, read:

> To our Lady the Queen:
> Most like those impious giants he contends,
> Who in Jove's empire his own sway extends,
> Where the Assyrian ruled, where Persian reigned,
> There Oxus' flood and Indus' there restrained.
> In vain the Macedonian sheds his tears
> Because to him too small the world appears.
> Even to Rome a watery bound was set
> Where Danube here and Tigris here she met.
> But wider the blue wave to Britain bows
> Where'er the breezes waft her mighty prows
> ANNA the sea's bright queen, Jove's ally thou,
> Don thou thine armour, crown thy royal brow,
> Pallas herself who grants thee all her aid,
> All men will think thee, in her image made.

He also praised the unknown inventor of the compass or nautical box:

> Him who first taught with magnetism to imbue
> The iron: and the ocean's watery waves
> Made clear to ships erst doubting: him who linked
> Shores, till his time far sundered, and by wind
> Brought mutual products to remotest lands:
> A thankless day, a heedless age have hid.
> No mightly name survives him, being dead.

Hope not to wrest thy fame from Stygian shades,
Nor seek to win thy ashes honours due.
And yet—to know within thy secret heart
A skill surpassing common mortals, to have blessed
The life of far-off grandsons, is not this
Itself the Elysian fields, the shining crown?

These long-gone verses were printed on charts for mariners and carried across oceans until the turn of the next century. At the time of its early publication, Halley had encouraged additional contributions to improve his chart. "All knowing mariners are desired to lend their assistance and information, towards the perfecting of this Useful Work. And if by undoubted Observations it be found in any part defective, the Notes of it will be received with all grateful Acknowledgement, and the Chart Corrected accordingly."

The later charts apparently also included data from other seamen. It was a consummate display of international scientific process: collaborative data gathering, with results open to criticism and ready to be amended by future improvement.

From existing copies of various editions, historians note that it was reprinted frequently with minor changes. No doubt the wider dissemination of information through increased trade helped advance science and navigation, including many of Halley's contributions. The utter volume of reprints and revisions seems to indicate the charts were unquestionably a practical success. But the constantly changing magnetic field of Earth also pushed revisions.

Whenever supporting evidence was presented to Halley or the Royal Society, he updated the charts. Other researchers frequently submitted new data, as evidenced by the steady stream of reports in the society's correspondence. But wide calls for data from other mariners were only heeded now and again. In a 1714 paper, written a year after he was honored by election to secretary of the Royal Society and assumed control of the *Philosophical Transactions,* which heralded its Newtonian era, Halley mentioned some recent observations of magnetic variation in part of South America by the French. In the same paper he also documented the naming of the Falkland Islands by a Captain Strong, who came across them while searching the South Seas for the hull plate from a wreck.

Printers in France and the Netherlands also published Halley's world chart. Around 1740 the Dutch East India Company adopted the Halley charts for its ships with the intent to use them to estimate longitude from the variation.

The need for accurate variation charts would continue for some time, and Halley's trend-setting isogonic charts would continue to be updated into the 19th century. To steer a ship properly, the helmsman required corrected courses derived from the chart. The continual shifting of the Earth's magnetic variation, however, rendered the ever-popular maps almost obsolete as soon as they were published, likely reducing their practical value, some scholars purport.

For determining longitude, Halley's method did not work for the most part. It proved useful for locations where the isogonic lines run parallel to a coast, such as that of southern Africa (at least at that period in time until the Earth's field shifted again). Halley was well aware of the shortcomings of his approach. It was impractical in most locations. Moreover, he knew that the points where it did work were temporal. "There is a perpetual, tho' slow Change in the Variation almost everywhere, which will make it necessary in time to alter the whole system," he acknowledged.

Halley's work confirmed the concept of secular or temporal variation, which he had observed as a schoolboy in 1672 and written about in 1683. His ideas on the subject expanded on Henry Gellibrand's 1635 discovery that magnetic declination changes with time.

"There is yet a further difficulty," Halley wrote, "which is the change of the variation, one of the discoveries of the last Century; which shows that it will require some Hundreds of years to establish a complete doctrine of the Magnetical system." With such a timeline in mind, Halley was clearly not expecting a revolutionary theory to emerge right after his *Paramore* mission.

Even after Halley published his charts that added to his luster, Flamsteed still held a grudge. Pursuing an ongoing investigation of Halley's 1683 submission to the Royal Society, Flamsteed asserted in a 1702 letter to Christopher Wren that, "I have some papers in my hands that prove him guilty of disingenuous practices."

The dispute remains unresolved by today's historians. Yet to many, Halley's 1692 paper on geomagnetism, which was clearly his own, rendered Flamsteed's charges irrelevant. It fleshed out his hypothesis of the detailed structure of Earth's interior. Very likely its analytical bent was strongly influenced by the style of Newton's *Principia*. In the second paper, Halley actually applied the physical limits spelled out by Newton to his postulate.

Regardless of the origin of some of the ideas in his first paper on geomagnetism, Halley's mission aboard the *Paramore* significantly contributed to what we now know as geophysics—more so than his basic meteorological chart and other previous innovations.

And Halley prospered in good company. Not only geophysics, but many scholarly disciplines we still recognize today were seeded in this period. For example, the roots of taxonomy can be traced to this time when methods for classifying animals and plants were first hotly debated. The work of John Ray, dubbed by Anglophiles the English Aristotle, was especially influential in establishing a classification system. Meanwhile, practicing as a physician, Hans Sloane, still secretary of the Royal Society, was well on his way to establishing himself as a great collector and cataloger of the living world, his eclectic natural treasures to form the basis of the British Museum. Profound insights about today's world can indeed be gained from Halley's.

In London in 1724 George Graham, although primarily an instrument maker, noticed that the compass needle every now and then veered off at a small angle for maybe a day or two. The effect was not only local. At the same time in Uppsala, the Swede Anders Celsius observed the same thing. A century later the effect was found to be worldwide, and Alexander von Humboldt would call such events magnetic storms.

Throughout this period, experimenters were really aware of only the existence of permanent magnetism, the type exhibited by magnetized iron or lodestones. On the surface the magnetic force due to the magnetic pole at the end of a magnet seemed a bit like gravity (or an electric force), which decreased in proportion to the inverse square of the distance from the pole. In 1777 in France, Charles Coulomb confirmed this inverse square relationship by experimenting with a magnetic needle suspended on a twistable spring. Magnetic detectors would be based on this instrument, which Coulomb invented, for almost two centuries.

Basic tenets of Halley's magnetic hypothesis would resurface over the next centuries, and later researchers would propose ideas similar to his. Christopher Hansteen, for one, a Norwegian physicist and astronomer who led an expedition to Siberia to study geomagnetism, put forth a four-pole theory in the early 19th century. And even a leading contemporary theory about Earth's magnetic field suggests, like Halley's, that the solid inner core might rotate at a different rate than outer regions. However, knowing that the Earth is not a permanent magnet, the theorists proposed an entirely novel process. (This notion was debunked when scientists recognized that materials lose their magnetic properties at high temperatures. The Curie temperature, as it came to be called, is about 770 degrees Celsius for iron.)

It would be more than a century after Halley's death when German mathematician and astronomer Karl Friedrich Gauss introduced improved techniques for observing and analyzing Earth's magnetic field. For example, in 1832 he devised the magnetometer, essentially a permanent magnet suspended horizontally by a gold wire, to measure the strength and direction of magnetic fields. In 1840, Gauss improved the technique by attaching a mirror to the suspended compass magnet. This enabled the angular motion of the compass to be accurately detected from a distance when a light source was bounced off the mirror and tracked. Another century would pass after that before Sidney Chapman and Julius Bartels would move the field into the modern age.

But in June 1701 the *Paramore* still had another mission in her. Halley wasn't prepared to relinquish his command just yet, and the monarchy was willing to oblige him one more time.

THIRD VOYAGE: 1701

are likewise to take the true bearings of the principal headlands on the
English coast one from another, and to continue the Meridian as often as
conveniently may be from side to side of the Channel in order to lay
down both coasts truly against one another.

What made tides ebb and flow had been pondered by such figures as Leonardo da Vinci. The 15th- and 16th-century Renaissance
man realized that the tides were linked to the Moon but had no in-
kling as to how. A century later Galileo, too, would take a stab at the
problem with no more success.

On the open sea, the effects of tides are negligible. But close to
land, tides and currents are key for navigating or for winning naval
engagements. By Halley's day, so much shipping passed through the
English Channel that any improvement in its navigation would have
reaped big rewards.

Although many great philosophers of the day did not fail to recognize the Moon's involvement, they, like da Vinci and Galileo before
them, had little understanding of how or why. To find answers to this
quandary, the leading minds sought to collect as much data as possible and then see what hypotheses might develop. Tides had been
explained theoretically only shortly before Halley's voyage with the
publication of Newton's *Principia*. Newton had solved the leading
part of the mystery revealing how the Moon's gravity was implicated.
As Newton's notions of gravity were yet to be widely applied, Halley
significantly contributed to the dialogue on how the Moon influences
tides as well as the *Principia's* practical use.

Before the third voyage Halley had apparently once again subtly
tried to ditch the *Paramore* for a better ship:

> That if their Lordships shall think it fitting to have an exact account of
> the Course of the Tides and about the Coast of England, so taken as one
> view to represent the whole . . . there be provided a small vessel such as
> their Lordships shall think proper, with all convenient speed, on board of
> which such an account of the tides may be taken, as their Lordships shall

direct; for which service their Lordships most obedient servant humbly offers himself.

But his desire to complete the mission during the summer overtook such fancy. Three days later on April 26, 1701, Halley requested that he depart immediately "in order to get the Paramour Pink manned with such complement as their Lordships shall think fitting." The Admiralty outfitted him with a pair of small boats, two extra cables, an additional anchor, and guns, as Halley specified, from the Tower of London, which housed the monarchy's chief armory.

The *Paramore* wouldn't sail until early June, however, because Halley had trouble manning his pink despite the favorable terms then offered by the Navy. Perhaps they could sense the danger of the spy component. Or it could have merely been the pay and labor. At the time of his third mission, the royal bounty for able seamen was 30 shillings and 25 shillings for ordinary seamen. Even combined with regular wages, this compensation wasn't competitive with that of private merchants.

Richard Pinfold was the only crew member besides Halley who signed up and sailed on all three voyages. Pinfold was promoted from a captain's servant to a captain's clerke for the third trek, according to the wages book. They set sail without a full crew, short a few hands. On the way down the Thames from Deptford, Halley eventually procured four men from Rear Admiral Munden, whose 60-gun *Plymouth* was then in the Downs.

The tidal survey presented different challenges for Halley's crew than his seagoing missions. The physical work was just as arduous but in different ways. Perpetually weighing and anchoring is extremely labor intensive, as is rowing a pinnace in circles to chase a tide. The ship still needed to be cleaned but not overhauled like at sea. Day-to-day dangers were not as great, as the shoreline was usually within sight. And they started exclusively on the friendly English side.

Daily, Halley recorded high and low water, the set and strength of the flood and tides, and stages of the Moon. He noted in particular any anomalies in the flow, including half tides. He collected data on the "sand, shoals, depths of water, and anchorage" as well as the tidal flow and currents.

A month into the voyage, Halley once again crossed paths with Sir Clowdisley Shovell at Spithead, this time without incident. Shovell, who had overseen Halley's court-martial proceeding in London and still sported his favorite emerald ring on his finger, was then captaining a flagship called the *Triumph*.

It was not long before Halley decided to attempt a foray on the French side of the channel. On July 11 the *Paramore* had reached Aldernay, one of the small channel islands off the coast of Britanny. His log of the third voyage details a typical day:

> This morning early, the ebb slackening, I weighed and stood in [headed] for the Island, but having little wind, the flood came so quick, that I was obliged to anchor again. Aldernay bearing WSW and Cape Jouber SSE in the middle of the Race. But even here the flood came without the Island. The Tide ran with great violence between 8 and 9 this morning, and by the log better than five knots. About 11 resolving to pass the race I got under sail with the last of the flood and at twelve Aldernay baring due west I observed the latitude 49 degrees 47 minutes. As soon as we were through the race and had gotten the French land NE of us the strong tide abated and the ebb sat SSW between Sark and Jersey, in the afternoon the wind came to W and WSW so that we could not lie better than South, and withal it began to blow fresh so I resolved to put in to Jersey.

Everything went well. The next day back in Jersey on English soil Halley would acquire a 24th man, a pilot named Peter St. Croix. (He never filled the 25th spot authorized by the Lords.)

Like many a seaman, Halley harbored a long-standing interest in tides. In the summer of 1678, a Mr. Francis Davenport sent a letter to the secretary of the Royal Society. The letter detailed mysterious tides at the port of Batsha in the Gulf of Tonkin in southeastern China that ebbed and flowed typically only once a day.

Davenport observed the tides mainly to help large East Indiamen secure passage across the river at the bar. He surmised that the irregular tides might be attributed to a sandbank as well as assorted inlets in Tonkin bay. The tides were especially erratic during the monsoon season. He supposed that the combination of seasonal changes in the Moon's motion and the monsoons caused the seeming randomness of the tides. He also enclosed a thorough data set of times, dates, and variation of the tides.

As a newly appointed Royal Society fellow, Halley jumped on the problem of the Tonkin tides. He identified a complex pattern behind the phenomenon and recognized that the daily high tide resulted from the rising Moon during the first half of the month and the setting Moon during the other half. He ruled out the monsoons as a factor and instead hypothesized that the bay of Tonkin's tidal range was proportional to the position of the Moon with respect to where the Sun's path crosses the celestial equator.

In his paper, which was published in 1684 and formally established the Moon's influence with an equation, Halley modestly commented on how little the world knew of tides, including those off British shores "of which we have had so long experience." But the completion of Newton's *Principia* would change such perspectives. Newton had realized that to strengthen his masterpiece he needed to add a general theory of the Moon's motion, which would contribute to explaining the mystery of the tides.

Flamsteed, who viewed himself as the authority on tidal prediction, was displeased with Halley and Newton's endeavors on the subject, but he gave Newton his data on lunar motion albeit grudgingly. Many believe it was Flamsteed and Halley's discourse over tides that led to the rift between the two astronomers.

Flamsteed had published charts on the tides at London. The then royal astronomer wrote that "there is every where about England, the same difference betwixt the spring [high] and neap [low] tides that is here observed in the river Thames." However, he was wrong. And

Halley was quick to note that. The tides in a river are not necessarily related to tides at coastal ports, he asserted. "But it is hence found that the said tables are not applicable to sea ports, where there is not the same reason for anticipation of the neap tides upon the quarter moons." Not to mention that in the big picture, tides in shallow seas and estuaries are more complicated than those in the open ocean.

Deftly sidestepping Flamsteed's jealousy, Halley went ahead and summarized tidal theory as explained by Newton's laws and presented the work to King James II, who was also known for his seamanship, along with a copy of Newton's *Principia*. Halley's explanation was intended for readers curious about tidal phenomena but not up to the task of understanding the high-level math entailed in Newton's treatise. Halley's contribution was soon added to the key manuals used to teach the art of navigation to seamen.

In Book III, the final book, of his *Principia*, Newton explained the anomalous tides in the port of Batsha in the Gulf of Tonkin, a feat Halley had previously believed impossible. "The whole appearance of these strange Tides, is without any forcing naturally deduced from these Principles, and is a great Argument of the certainty of the whole Theory," Halley exclaimed. It was Newton's attention to the tides that expedited the general acceptance of his theory of universal gravitation, most scholars contend.

Exposing his talent for science exposition, Halley wrote: ". . . being sensible of the little leisure which care of the Public leaves to Princes, I believed it necessary to present with the Book a short Extract of Solution of the Cause of the Tides in the Ocean. A thing frequently attempted but till now without success. Whereby Your Majesty may judge of the rest of the Performances of the Author." In this way it was Halley, who after Newton was the person most familiar with his *Principia*, who made the work directly accessible to those who weren't up to speed with Newton's mathematics.

In his letter presenting the polished *Principia* to James II, Halley summarized Newton's concept of gravity and how it regulated the motion of planets, comets, and the Moon. According to Halley,

Newton attributed the tides to the "decrease of gravity from the contrary attraction of the Sun and Moon, whereby the water being less pressed rises where they are vertical, and subsides when they are in the horizon."

In simpler terms, the way the Moon pulls various portions of the Earth differs. If Earth is viewed as a crust completely covered by water, the Moon pulls on the layer of water as it orbits Earth. The Moon pulls on the oceans on the side of Earth facing it more strongly than the remainder of the planet, because they are closest to it, causing the water to bulge toward the Moon—or a high tide. Meanwhile on the opposite side of Earth, the crust is closer to the Moon than the ocean. The Moon pulls the crust away from the deep oceans also inducing a high tide on the far side. As Earth rotates on its axis, each location on Earth will experience both tidal bulges. The areas of high water levels are high tides, and the areas of low levels are low tides.

But the dynamics are further complicated by the Sun. While the Moon orbits Earth, the duo, together, also orbit the Sun. When the Sun and the Moon line up, their gravitational forces combine to cause very high and very low tides, which are called spring tides. When the Sun and the Moon are out of kilter, the gravitational forces cancel each other out, causing neap tides, which are not as markedly high and low. Newton's *Principia* went on to elucidate why tides vary from day to day as well as place to place with latitude, as gravitational effects shift with the constantly changing interactions between Earth, the Moon, and the Sun.

Halley also explained to the king in greater detail Newton's explanation of the strange tides at Tonkin, where only one tide occurs each day. The second tide is effectively cancelled because two tidal streams run from the South China Sea into the Gulf of Tonkin. They are out of phase with each other by six hours due to geographical obstacles in one of the streams.

Newton properly credited Flamsteed with supplying the lunar observations that made the later edition of the *Principia* possible: "All the world knows that I make no observations myself," he wrote in a

February 1695 letter thanking Flamsteed. Apparently Newton was in Halley's camp on the issue of sharing scientific data.

Before Halley's sea adventures, Newton also helped him develop his theory of the secular acceleration of the Moon, which said that the satellite actually increases its velocity with respect to Earth. According to the Royal Society *Journal Book*, Newton had surmised "that the bulk of the Earth did grow and increase . . . by the perpetual accession of new particles attracted out of the ether by its gravitating power," and Halley proposed that "this increase of the moles of the Earth would occasion an acceleration of the Moon's motion, she being at this time attracted by a stronger [centripetal force] than in remote ages." Halley demonstrated this phenomenon by matching up predicted times of ancient eclipses of the Moon with recorded ones.

Shortly thereafter, in 1697, Halley lectured the Royal Society on the propagation of tides near the British Isles. And he published his later tidal paper in the *Philosophical Transactions.*

At the time of this third voyage, tensions with Spain were mounting. The part of Halley's mission that entailed intelligence collection on the English Channel was a well-guarded secret. Martin Folkes, who would assume the presidency of the Royal Society in 1741, for one, asserted that Halley's channel survey was merely a cover for his clandestine mission. Halley was dispatched to gather intelligence about the defensive strength of the French near channel ports.

What's more, to Halley's original instructions the Admiralty added the following warning about taking liberties in publishing any data he gathered while on the mission: "And in case during your being employed on the Service, any other Matters may Occur unto you the observing and Publishing whereof may tend towards the Security of the Navigation of the Subjects of his Majesty or other Princes trading into the Channel you are to be very careful in taking notice thereof."

Unlike Newton, Halley had few reservations about applying science to improve military capabilities. In fact, Halley had become an expert in many aspects of weaponry and combat.

Another early practical application of Newton's gravity work was determining the variable rainbow arc of a cannonball or other projectile to make it more deadly accurate. In 1686 Halley coauthored a paper on gunnery. It was titled "A Discourse Concerning Gravity, and Its Properties, Wherein the Descent of Heavy Bodies, and the Motions of Projects [projectiles] Is Briefly, but Fully Handled; Together with the Solution of a Problem of Great Use in Gunnery."

The trajectory problem was a long-standing one that required Newton's calculus to be properly solved. Halley calculated how a projectile should travel through the air under Newton's new theory of gravity. He demonstrated how to set up a mortar to hit a given target no matter its starting point by fiddling with its elevation. Halley also recognized the value of developing standard mortars, ejectable bombs, and gunpowder charges to increase targeting accuracy.

Five years later he also authored an unpublished paper on how high bullets should be shot to reach their intended targets. In some of his other papers he wrote on such practical topics as how to balance the weight of guns on a ship's deck to minimize strain on the hull.

As Halley continued the channel survey, whenever he went into port whether for provisions or due to inclement weather, he and his crew would hear rumors that war was indeed on the horizon. But he refused to let stormy skies of either persuasion interfere with his mission. Halley's third voyage may have been short on exotic glamour compared to his first and second voyages, but it would bear practical fruits for channel travelers.

In a letter dated September 13, Halley wrote to Burchett reporting the success of his mission. He knew the Admiralty would be pleased with his coastal survey and his tidal observations. What is more, he mentioned that he may have discovered some general principles along with his useful findings.

On Halley's return to London in mid-October, the *Paramore* was

laid up at Deptford and her wages paid off. Halley's surgeon William
Erles received about 24 pounds and his boatswain Richard Price
roughly half that or 12 pounds. Able men were paid six pounds apiece
for more than a five-month-long mission. Payment of the bulk of
Halley's 142-pound salary ($143,000 in U.S. currency today) was de-
ferred. He received a mere three shillings at the payoff.

In wages the total amount doled out for all three of Halley's voyages barely exceeded 1,000 pounds ($1 million in 2005). The figure
includes compensation of all three crews—roughly 100 men—for
their respective duty periods.

By now Halley had become something of a celebrity, an English
darling with his popularity soaring. Queen Anne, perhaps impressed
by his poem to her, gave him a bonus of 200 pounds on top of his
wages "for his extraordinary pains and care he lately took" in surveying the tides and the channel. His resulting chart, "A New and Correct
Chart of the Channel Between England and France . . . with the Flow-
ing of the Tides and the Setting of the Current," represented another
precedent. It is considered the first true tidal chart.

Once again, Halley was thinking ahead of his time. No one would
conduct a similar survey for a century or publish a tidal chart for
another large body of water for 150 years. His chart would be widely
used by mariners throughout the 18th century. Along with detailing
the shoals, anchorages, depths, and coastlines like other maps of the
day, Halley's map offered two key improvements.

The first was a formula for estimating the height of water in the
channel at specific sites according to the Moon's position. They are
demarked by Roman numerals on the map. The second was a new
method for surveying the coast (see Appendix). Called the resection
method, it relied on fixing horizontal angles by the Sun to achieve
greater accuracy than the conventional practice of using a magnetic
compass. "This is a very easy and expeditious way for putting down
the soundings in Sea Charts in their proper places, and may be practiced in a ship under sail," Halley advised.

Nothing is known of the results of the spy mission. No records survive. Either Halley failed to collect anything of value or he succeeded brilliantly in keeping all of it secret.

Halley was due to get his land legs back and return to the ranks of civilian life. He set about publishing the comet work, which he had undertaken a decade earlier, before more important matters with the *Paramore* interrupted him.

CHAPTER 11

QUEEN ANNE'S PATRONAGE

W hen King William died without an heir in 1702, Queen Anne rose to the throne and Halley had his third bene-factor within a decade. Since the Glorious Revolution, she had loyally sided with William and her sister Mary.

Anne's first year as queen saw the War of the Spanish Succession break out. And suddenly Halley's extensive channel data jumped in significance. The conflict, sometimes referred to as Queen Anne's War, was fought over Europe's balance of power and domination of trade. Louis XIV had aspired to control Spanish possessions in northern Europe and America. England, allied again with the United Provinces (now the Netherlands) and the Austria-Hungary Empire (as it had been in the Nine Years' War against the French), wanted to check the Sun King's expansionist ambitions.

As the Royal Navy prepared for action, the *Paramore* was retooled and armed with heavier guns. The sound of unfamiliar footsteps would resound on her deck and in her captain's quarters. As soon as the war broke out with France, the ship served as a ketch for bombs in

Sir George Rook's squadron under the captainship of Unton Deering and later Josiah Mighells.

Queen Anne appointed Halley in 1702 and 1703 to diplomatic missions. He visited Vienna and Hanover, adding to his royal contacts. Emperor Leopold and Prince Eugene of Savoy and Istria received him. And in Hanover he dined with England's electoral prince, later King George II.

After returning to London from Vienna to advise the Emperor on forts along the northern shores of the Adriatic, Halley was charged with overseeing construction of the defenses of the channel he laid out in his survey. The nature of this project lends credence to innuendo that Halley was clearly spying and not merely gathering scientific data on his channel survey. And as further evidence, he was openly paid for his work in Vienna "out of secret service" funds.

Even though the war was still being fought by 1705, Halley's chart of the English Channel was published in Amsterdam along with a printed explanation in Dutch and French. After a stint in the Mediterranean, the *Paramore* returned once again to Deptford for repairs. In 1706 she was sold to a Captain John Constable for a mere 122 pounds—less than his wife's dowry—or roughly 64,000 pounds today.

With three voyages under his belt and service as a government agent, Halley's name was put forward to fill the prestigious Savilian Chair of Geometry at Oxford. "Dr. Wallis is dead. Mr. Halley expects his place. He now talks, swears, and drinks brandy like a sea captain, so that I much fear his own ill-behavior will deprive him of the vacancy," was the searing gossip Flamsteed sent to his assistant Abraham Sharpe in December 1703.

But despite Flamsteed's unsurprising opposition, Halley was unanimously elected, his popularity at a new high. After all, a month earlier he was elected to the council of the Royal Society with more votes than even Newton collected. Newton, naturally, had won the Royal Society presidency that year.

As Halley settled into academic life at Oxford, he built a small observatory atop his own house (which still stands today). The seeing may not have been of the quality that he enjoyed at St. Helena, but at least he could continue his astronomical observations uninterrupted. He also continued to observe on his many visits to London. By 1710, Oxford had added an honorary doctorate to his laurels.

Years before Halley's voyages, Newton had begun working on the theory of the Moon's motion. As early as 1694, he had turned to Flamsteed for help. He wanted the royal astronomer to publish his lunar observations for the second edition of the *Principia*. Initially, Flamsteed supplied some of the desired data, but then their relationship deteriorated. Now a decade later, with his mounting influence, Newton pushed to have Flamsteed's observations and star catalog (to be titled *Historia Coelestis*) published once again. After prompting from Newton, Flamsteed secured funding for publishing the observations from Queen Anne's consort, Prince George of Denmark, who also appointed five Royal Society reviewers to guide the process. More than 100 pages of the estimated 1,450 pages were printed and hundreds more had been reviewed by Flamsteed when the prince suddenly died in 1708. Progress again ceased on the publication.

Likely under pressure in part from Newton, Queen Anne authorized a handful of Royal Society council members to oversee Flamsteed's activities at the observatory, infuriating Flamsteed. They were to purchase and keep up the facility's instruments, which were technically Flamsteed's personal property, and ensure that his results were published annually. Adding insult to injury, the queen appointed Halley—without even consulting Flamsteed—to edit his prized *Historia Coelestis*.

Halley's preface, finally published in 1712 as part of his edition of *Historia Coelestis*, was duly insulting: "Flamsteed had now enjoyed the title of Astronomer Royal for nearly 30 years but still nothing had yet emerged from the observatory to justify all the equipment and expense, so that he seemed, so far, to have worked only for himself or at any rate for a few of his friends, even if it was generally accepted that all these years had not been wasted and that the Greenwich papers had grown into no small a pile." While Newton and Halley's dissatisfaction with Flamsteed was understandable, their equally egotistical conduct reasonably irked Flamsteed. The friction between these rivals didn't end there.

THE TREATY OF UTRECHT BROUGHT an end to Queen Anne's war in 1713. For Britain it was a success. England had gained control of Newfoundland and most of its unusually prolific fishing waters. The newly fashioned peace would permit the monarchy to focus once again on diplomatic means of improving trade.

In 1714, testimony from both Halley and the increasingly irascible Sir Isaac Newton helped pass the Longitude Act, the world's first scientific legislation. On July 20, Queen Anne gave the nod to the bill, which was encouraged by William Whiston, who had succeeded Newton at Cambridge as Lucasian professor, and a fellow mathematician named Humphrey Dutton. By now even the more conservative members of Parliament had faith that 20,000 pounds could spark native ingenuity into solving a complex problem. The implication was that society needn't wait passively when confronted with pressing scientific challenges.

Whiston and Dutton aspired to win the prize themselves by stationing ships at steady intervals on trade routes and then having their captains fire star shells at midnight that would all explode at the same height—6,440 feet—above the water. Then the longitude of a passing sailing vessel could be calculated by determining the distance to a fixed signal ship from the time lag between the scheduled explosion (or actual sighting of the fireball if it weren't overcast) and the time the signal was heard aboard their ship.

Though it would have been expensive and impractical to implement, their approach was no more or less off the wall than the multitude of others out there, ranging from timing the effect of a magic potion released in London on injured dogs that were carried aboard ships to painting longitude meridians in the evening sky. Newton pompously declared a chronometer—what would prove one of the viable solutions—incapable of solving the longitude conundrum; Halley was wise enough to consider all options and even supported early on the eventual winner, whom almost everyone had doubted: John Harrison and his mechanical marine clock, the H-4. For example, Halley was among the few Royal Society members who advocated for its trial when Harrison completed the timepiece in 1735. In truth, probably the best chance for solving longitude at this point was using several of the methods in combination to cancel out each other's problems.

Halley's geomagnetic survey, which proved that there was not much correlation between magnetism and lines of longitude, was also influential in prompting the Longitude Act. Conversely, if his magnetic mission had failed, the resulting lack of confidence in public ventures in science might have undercut Parliament-sanctioned funding of bigger and better endeavors in the future.

The Longitude Act was passed just 12 days before Queen Anne's death on August 1, 1714, before reaching the age of 50. She left no heir either, bringing a close to the House of Stuart. Her husband, Prince George, had died six years earlier and her only son William, Duke of Gloucester, died at age 11. He was the only child of 17 pregnancies to live past infancy.

Anne had succeeded in strengthening the British domain. Her reign oversaw the 1707 Act of Union, which created Great Britain by truly uniting England and Scotland under the governance of one parliament: For the first time, government of the whole country was centered in London (Wales had been joined to England since the 13th century). In accordance with the Act of Settlement, the elector of Hanover would succeed Anne.

Halley reportedly took an oath of allegiance at Westminster Abbey to the new King George I, who came to the throne on Queen Anne's death in 1714. His ascension transformed the government from Tory to Whig, which favored Flamsteed once again. And Parliament now held the authority of the oligarchy and would gradually become the dominant power in England. The royal prerogative was left effectively intact. Anne would be the last British monarch to ever veto an act of Parliament.

Flamsteed died at age 73 some five years after passage of the Longitude Act and the change to the House of Hanover. Halley finally could claim the astronomical throne to which in the minds of many he was long overdue. From his perch as astronomer royal, Flamsteed can be credited for helping spawn the business of scientific instruments and chronometry in England and the resulting improvements in observation and measurement. King George officially named Halley to replace Flamsteed on February 9, 1720.

WHILE PINING FOR THIS POST, Halley had made serious contributions to mathematics. He had published more than half a dozen papers on pure mathematics between 1687 and 1720. They ran the gamut from computing logarithms and trigonometric functions to delimiting the roots of equations. His writings on applied mathematics from his mortality tables to his gunnery trajections were also of note. Yet Halley had been preparing for this next step all his life.

At age 64, Halley had landed the long-sought-after post of astronomer royal. And—always the explorer—with his adventurous spirit still very much alive, he boldly wrote a blueprint for a 19-year-long series of observations of that tidal force and eternal inspiration for poets and scientists: the luminous Moon.

Indeed as England's second astronomer royal, Halley had joined the ranks of the world's scientific elite. At this point in his life, only a score or so of men were actually paid as scientists in Europe, including the astronomers at Greenwich, Paris, Berlin, Oxford, Cambridge, and the like; the curators of Padua's and Leiden's botanical gardens; and probably a couple of anatomists and chemists at universities and other institutions. Other academics were paid as mathematicians as well.

Flamsteed had believed "that observations of the Moon's distances from fixed stars were the most proper expedient for the discovery of it [the longitude]." Halley himself was committed to the lunar distance solution to the longitude problem. The method requires that the position of the Moon at any given moment, relative to the Sun or other stars, be predicted accurately.

Halley had conceived of using the cycles of the Moon to determine longitude in 1682, years before Newton had developed his theory outlined in the *Principia,* which made such a scheme seem possible. (Of course, Nuremberg's Johann Werner first put forth the method in 1514.) Halley also considered using lunar and solar eclipses as timepieces in the sky.

Meanwhile, the Italian astronomer Cassini in Paris was more partial to using another celestial clock, that of Jupiter's satellites, to find longitude. Galileo first proposed the idea when he discovered the planet's four satellites. Like Earth's moon, Jupiter's moons are independent of Earth's rotation. Cassini constructed elaborate tables of the satellites' motions. Robert Hooke also engaged in making tables of Jupiter's occultations.

Halley, however, criticized Cassini for rejecting his Danish colleague Olaus Roemer's theory of light, which explained irregularities in the times of the eclipses. Light, contrary to accepted beliefs, had a finite velocity and did not arrive instantaneously, Roemer explained in a paper later published in the *Philosophical Transactions.* (Like Cassini and Huygens, Roemer was invited to Paris to join the French academicians, who received pensions from the king.) Halley contended there were big errors in Cassini's tables because he ignored Roemer's findings. Halley attempted to correct the tables for London. But the errors even for Jupiter's first satellite, which was the least affected by the ramifications of the fixed speed of light, were as large as three minutes or 45 degrees of longitude. Halley used components in Newton's *Principia* to explain even the smaller anomalies. For example, Halley supposed that the nature of the gravitational attraction of Jupiter's equatorial bulge could cause minor irregularities in the satellites' orbits. (In Book III of the *Principia,* Newton demonstrated that rotating bodies like Earth and Jupiter were spheroids.)

Just the same, when Halley reached the New World on his voyage, he made telescopic observations and used Cassini's tables to attempt to improve his dead-reckoning estimates of longitude while crossing the Atlantic. In places they were more than 400 miles in error. In the absence of a fleshed-out theory of the Moon, Halley recognized that the eclipses of Jupiter's first satellite were the best-available clock for finding longitude on land. Improvements on Cassini's tables weren't made until 1719.

ALMOST POETICALLY, HALLEY HAD started and would culminate his career at the Royal Observatory in Greenwich not far from his place of birth. At that historic site the prime meridian, the arbitrary line of zero degrees longitude, would one day be universally agreed upon. Atop a rolling green hill, the enchanting red-brick building with its observing domes majestically overlooks the Thames and gives a full view of the ingoing and outgoing shipping traffic from London.

Reality would subdue Greenwich's charm a bit, however, when Halley moved into the observatory and adjoining house in March 1720. Flamsteed's widow had taken all of her husband's instruments from the premises, as he'd purchased many of them himself. His instruments not only were valuable but were also symbols of mathematical and professional prowess. Without the proper tools, Halley wouldn't make his first observation until October 1721. After muddling through with his instruments, Halley secured a 500-pound grant from the government to reequip the observatory in 1724. His first purchase: a mural quadrant, with an eight-foot radius, to measure zenith distances of stars moving toward the meridian in order to determine latitude. Crafted by George Graham, one of the leading instrument makers of the early 18th century, the device consisted of a circular border that forms a 90-degree arc attached to an arm that is pointed at the celestial body. Jonathan Sisson had devised an innovative design that afforded more accuracy than ever before. The model would help spur England's instrument export business in the coming decades.

With restoration of the observatory under way, Halley concentrated on his scheme to observe the Moon daily over its entire 18-year cycle or saros. From these data he planned to make empirical tables of the Moon's motion.

A year later, in 1725, Flamsteed's widow, abetted by his assistants, published Flamsteed's own version of *Historia Coelestis,* which cataloged nearly 3,000 stars, and several years later an accompanying star atlas that displayed his data graphically.

After George's son succeeded the king in 1727, his wife Queen Catherine visited Greenwich and toured Halley's observatory there and was favorably impressed. When she heard that the salary of the Astronomer Royal was the same as that when Flamsteed was first appointed more than 50 years previously, she sought a raise for Halley. He, however, declined. "Pray your Majesty do no such thing, for if the salary should be increased it might become the object of emolument to place there some unqualified needy dependent, to ruin the institution." Catherine went ahead and obtained from her husband King George an annual grant for Halley's former services in the Royal Navy, half a captain's salary.

At the age of 82, Halley completed his observations of a full saros. Between 1722 and 1739, he eyed the Moon at meridian passage whenever possible. He bragged to the Royal Society in 1731: "With my own eye without assistant or interruption . . . 1500 observations of the Moon . . . more than Tycho, Hevelius, and Flamsteed had taken altogether. . . ."

Ironically, in 1727 Halley, like Flamsteed, had refused to publish his latest observations, so "that he might have more time to finish the theory he designs to build upon them, before others might take advantage of reaping the benefit of his labours." Flamsteed probably relished that fact from his grave.

PERHAPS HALLEY HAD THE 20,000-POUND longitude prize on his mind. The sum was the equivalent of more than 10 million pounds today. Clearly, his work confirmed the theoretical feasibility of the "method of lunars" as a way to solve the longitude problem. It is unclear whether he, as an ex officio member of the Board of Longitude, would have been eligible. But his detractors were quick to think the worst of him. Flamsteed had reportedly commented: "Raymer [his unflattering nickname for Halley] sets up for a finder of the longitude."

By as early as 1731, Halley had acquired enough data to establish the use of lunar observations for determining longitude at sea. Although at the equator the margin of error ranged up to nearly 70 miles, it was an improvement over existing methods. Around this time English astronomer James Bradley, Halley's eventual successor, discovered another celestial phenomenon that could also skew lunar observations, namely aberration, the ever-so-slight deflection of light through space. And then in 1748, he noticed yet another: nutation, the small oscillation of celestial bodies. He realized all past, present, and future lunar observations needed to be corrected for both.

Getting navigators to use the lunar distance approach was another matter, of course. It was not only tedious but time consuming, and better instruments, such as the Hadley reflecting quadrants, would move things along. With the help of the *Nautical Almanac* developed by Nevil Masklyne, who would become the fifth astronomer royal, the calculations became less unwieldly. First published in 1767, the *Almanac* detailed the distance to the Moon from various stars at short intervals from which Greenwich time could be computed.

Throughout the course of the approach's evolution, Halley's lunar observations would be heavily criticized for their inaccuracy. Although his Moon work clearly furthered Newtonian dynamics, the criticism must have held much truth for he would be the only royal astronomer whose lunar observations would never be published.

ALTHOUGH THE LUNAR DISTANCE METHOD would never be error-free even if determination of such astronomical benchmarks could be perfected, it would be employed until the early 19th century, when chronometers really came into their own. Or at any rate the lunar tables would be published in the *Almanac* through 1906, though no longer in use. One sage captain later reported that he had "not fallen in with a dozen men who had themselves taken lunar or had even seen them taken [since 1855]. . . . They are in fact as dead as Julius Caesar."

CHAPTER 12

BACK TO THAT COMET

Not long after Halley returned from his third voyage, he picked up the searing question that seemed to always re-appear in his life. His attention returned to that comet he witnessed a few times in 1682. The memory of its cascade of seem-ingly burning light held his intrigue, its ephemeral passage replaying in his mind as if on its own peculiar orbit within Halley's skull. He hypothesized the comet not only orbited the Sun but that it was the same comet that had been seen earlier in the century.

What had mesmerized and often terrified Earth-bound audiences over the ages now—if Newton was right—had to follow the laws of the universe. Under the corrected vision of this age, comets were not errant anomalies on haphazard voyages but merely another class of orbital body governed by the cosmic clockwork.

The reason cometary paths were so hard to track is that they were visible so briefly compared with such celestial bodies as planets or stars. Measurement was tricky. At the time, determining whether their paths formed an ellipse, a hyperbola, or a parabola was next to impossible visually because comets could be seen only when they were close to the Sun.

With Newton's calculus and his conception of gravity in hand, Halley had the tools to change this. Compiling every available bit of data, he took a fresh look at the past 200 years of comet observations by the world's leading stargazers, including the tenacious Tycho Brahe, Kepler, Hevelius, Cassini, Flamsteed, and more. Computing the paths of comets with their bright heads and glowing tails was no longer a futile exercise, he reckoned. "I think there can be nothing plainer than that comets do move in orbs about the Sun [that approach the] parabolic."

But the actual computations, which factor in the gravitational pull of not only Earth but also distant planets as the comet approaches, were highly complex. Not only was Halley concerned with the exact path the comets traveled but with the future path of their next trip around the Sun. He had to account for the perturbations caused by solar system giants Jupiter and Saturn and the precise position of Earth when the comets had been sighted or were to return.

For Halley it was the perfect astronomical puzzle. Over the next few years he completed the calculations of the orbits of some 24 comets. He described the endeavor as an "immense labour." He assumed the orbits were parabolic. (One in particular would prove an exception.) Among other comets, he looked at the comet of 1680, which he had seen in Paris on his European tour. He calculated its orbital period to be 575 years, similar to those being talked about in Paris. Given, in part, that the estimate was unverifiable in human timescales, he devoted most of his attention to the comet of 1682.

Halley had started calculating that particular comet's period even before his adventures aboard the *Paramore*. At a June 3, 1696, meeting of the Royal Society, according to its recording secretary:

> Halley produced the Elements of the Calculation of the Motion of the two Comets that appeared in the years 1607 and 1682, which are in all respects alike, as to the place of their Nodes and Perihelia, their Inclina-

tions to the plain of the ecliptic and their distances from the Sun; whence
he concluded it was highly probable not to say demonstrative, that these
were but one and the same Comet, having a Period of about 75 years."

But this was hardly fleshed-out. At the heart of the comet debate
was the question of whether the comets were primarily repelled from
the Sun or attracted by it by a seeming magnetic force. To prove
Kepler's inverse square law for planets, it didn't matter whether
the force that influenced their orbits was gravitational attraction or
magnetic action, for both operated on the so-called inverse square
principle. The planets were attracted toward the Sun by a force proportional to the inverse square of the distance between the planet
and the Sun. The chief difference between gravity and magnetism,
we have since learned, is that while gravity only attracts, magnetism
both attracts and repels.

It was Halley who had initially persuaded Newton himself that
the inverse square law of gravitation held true throughout the solar
system. After accepting this notion, it was simple to make the leap
that comets might be influenced by gravity to the extent that their
orbits were elongated. Kepler had rejected the notion that comets
could follow elliptical paths. He argued that only eternal figures like
planets could chase such orbits. Just how stretched out the shape of a
comet's orbital circuit might be remained a key question.

After much ado, Newton developed the first theory of cometary
motion as part of his *Principia*. His synthesis of universal gravitation
would then forever link comets with Earth.

On his return to London and reunion with his family after his
channel survey, Halley had analyzed the existing data further. Like his
world map, it relied on numbers from many intrepid observers, this
time across centuries. As he tackled one tedious calculation after another, Halley persuaded himself that the orbits were elliptical. From
the outset even Newton had favored the more open shape of a parabola, which could extend infinitely in theory, never to return along
the same path.

Increasingly, Halley became convinced the comets of 1531, 1607, and 1682 were the same celestial body that returned every 76 years. He was now prepared to predict that the comet would next visit in 1758 at Christmas time. The forecast was the first of its sort.

Halley was 49 years old in 1705 when he boldly presented his prediction and budding proof to the Royal Society. His *Synopsis of Cometary Astronomy*, touting it, also was published at Oxford that year. He knew the chances were slim to none that he'd live to see whether his prophecy would come true in 1758. But he confidently asked the younger generation of scientists who would likely live to see the day "to acknowledge that this was first observed by an Englishman."

Many doubted Halley's hypothesis. Over the years even Halley would hedge his prediction. In 1717 he wrote that the differences between the orbits of 1531, 1607, and 1682 "seemed to me a little too large." He noted that the variations in the successive periods were "much larger than those which we observe in the revolutions of any single planet, since one of these periods exceeds the other by more than a year." He also noted that "the inclination of the comet of 1682 is 22′ larger than that of the comet of 1607."

In 1717, while plumbing medieval accounts, Halley found evidence of comets arriving before the Renaissance that might well be additional early sightings of the comet that reappeared in 1682. They had occurred on Easter 1305, an unspecified month in 1380, and in June 1456.

At this time he also expanded the window of the comet's return to late 1758 or early 1759. "All is nothing but a light trial, and we leave the effort of making this matter deeper to those who survive until the event justifies our predictions."

ONCE ENSCONCED COMFORTABLY as Savilian Professor of Geometry in Oxford's storied halls, Halley seemed to be more candid once again about his beliefs. For one thing, publication of the second edition of Newton's *Principia* in 1713 likely burnished Halley's reputation and he probably felt more secure in his career. It was not long before he returned to the controversial quarry, the question of whether Earth's life span is limited. This quest would summon forth both of Halley's longtime passions—comets and geomagnetism. And once again the story features Isaac Newton and Robert Hooke.

At the time that Newton was crafting the *Principia*, Hooke was developing an "excellent System of Nature" of his own (which he presented in talks known as his Lectures on Earthquakes delivered to the Royal Society in 1686 through 1688). The persnickety Hooke was extremely gifted—even the unique genius of Newton could feel challenged.

Hooke's general theory put forward Earth as a dynamic, living body, one whose fertility changed over time. At his scheme's center were his ideas about the "Cause and Reason of the present Figure, Shape and Constitution of the Surface of the Body of the Earth, whether Sea or Land, as we now find it presented unto us under various and very irregular Forms and Fashions and constituted of very differing substances."

Embracing his own theories on earthquakes, volcanoes, continent formation, magnetic variation, and even catastrophic changes in species, Hooke's new history of the natural world, in brief, linked celestial mechanics to Earth's dynamics. Everything from the formation of continents to changes in species were tied to his new treatise on Earth's spatial orientation in relation to other planets and stars, as governed by the inverse square law, and were linked to Earth's shape.

In this way, both Hooke and Newton's philosophical systems were based on the inverse square law of gravitation, for which Hooke claimed priority but for which Newton provided the mathematical proof for the orbits of planets in his *Principia*. Hooke also wanted credit for all that was derived from the law, which was no small presumption. While their competing derivations basically offered an explanation of the same universe, the visions differed greatly in scope and substance. To Newton, Hooke's demands were excessive and offensive. To Hooke, Newton's *Principia* had stolen the thunder of his grand synthesis of the universe and even borrowed some of his hypothesis's core tenets, in particular his ideas about the inverse square law and planetary forms.

Hooke's system hinged on the idea that the axis of Earth's rotation underwent "certain slow progressive motion[s]." As a result, he asserted that Earth was an oblate spheroid. In his mind, his revelation connected the planet's spatial orientation to the rest of the solar system and illustrated his dynamic history of Earth.

Initially, some key figures like Royal Society fellow John Aubrey, a noted diarist, bought Hooke's system wholeheartedly: "This is the greatest discovery in nature, that ever was since the world's creation: it never was so much as hinted by any man before," he exclaimed in a letter. Halley and Newton apparently took Hooke's ideas seriously enough to race him to press.

Halley even publicly accepted some of Hooke's premises and credited him with suggesting that Earth was an oblate spheroid. He also wrote to John Wallis, then head of the Oxford Philosophical Society, that Newton "falls in with Mr. Hooke and makes the Earth of the shape of a compressed spheroid, whose shortest diameter is the axis, and determines the excess of the radius of the equator."

Hooke's theory kindled Halley's interests in other ways. Halley accepted Hooke's notion that Earth's axis of rotation changes. He doubted, however, whether the type of gradual flooding that would have occurred as a result of changes in Earth's rotational axis would have been extreme enough to wipe out entire species. Instead, Halley suggested in a 1686 paper that Earth's axis might undergo rapid shifts caused by "the powers that first impressed this whirling motion on the ball [Earth]" or, perhaps naturally, "by the causal shock of some transient body, such as a comet."

Still in the midst of editing the third book of Newton's *Principia,* which covered cometary motion, Halley in his 1686 paper became the first from the Newtonian school of thought to put forth, publicly at least, the idea that comets played important roles not only cosmologically but also on Earth.

After the debut of the *Principia*, it would be Newton's theories that were generally accepted and Hooke's that remained at least couched as but "conjecture, of what may be, without any evidence from observations," in the words of Oxford's Wallis, "too extravagant for us to admit." Perhaps once again, artful behind-the-scenes campaigning by both Halley and Newton may very well have contributed to the different reception of these rivals' works.

In a later paper published in 1692, Halley expanded this collision idea and his 1683 hypothesis that Earth's magnetic variation may be explained by four magnetic poles in its interior—the one Flamsteed had rudely alleged Halley borrowed from the unsung mathematician Peter Perkins. In a novel synthesis, Halley asserted that a cometary impact on Earth might convey different velocities on the inner and outer shells, which each held a pair of poles.

He suggested that some sort of fluid medium might separate the two shells or that there may even have been a significant air cavity between them—some 500 miles wide—perhaps as thick as the outer shell itself. He reckoned gravity could hold a concentric sphere within Earth much like the rings of Saturn are captured by gravity. (It was not then known that Saturn's rings rotate; Newton hadn't considered the phenomenon in his *Principia*.) Halley's speculation was based on Newton's erroneous designation in Book III that the Moon is denser than the Earth by a ratio of 9:5. (The actual mass ratio has since been found to be 1:81.) "I have adventured to make these subterranean orbs capable of being inhabited," Halley added to the eternal pleasure of future science fiction writers.

Perhaps amazingly, this notion of a hollow Earth still chimes with possibility today. While some scholars contend Halley borrowed from Hooke's earthquake lectures to develop his idea that Earth was comprised of shells, there is a stronger case to be made that his work was original and more directly stemmed from Newton's incorrect estimate of lunar mass. Moreover, in Hooke's view Earth's interior was comprised of layers like an onion. His model never mentioned a series of magnetic poles, air gaps, or subterranean globes.

Yet it was with prompting by Hooke and Newton that Halley brought his ideas on Earth's magnetism and comets in space full circle. He built on both Hooke's changeable axis theory and Newton's ideas about the role of comets in the origin of the universe, detailed in his *Principia*.

In 1694 before the Royal Society, Halley had expanded his thinking about Earth's internal structure further. He hypothesized that the impact of a comet might not only explain Earth's magnetic variation but might also have caused the flood behind the story of Noah's Ark. In this unpublished work, he also commented that a comet might have wrought the annihilation of a "former world," creating the "chaos out of whose ruins the present [world] might be formed." The assertion had cost him the chance of winning the Savilian Chair of Astronomy at Oxford only a few years before.

Halley apparently attempted to clear his name in the scientific community of the charge that he had denied Earth had a finite age. Strategically, he intentionally put a more sophisticated spin on this question in his subsequent writings. In a prominent 1715 paper, for example, he refuted "the notion of the Eternity of all Things, though perhaps the World may be found much older than many have hitherto imagined." His assertion was based on his observations of the saltiness of the seas and lakes, which placed a limit on the age of the Earth. He proposed that the age of the Earth is limited by the rate at which rivers carried salt from the sea into lakes without outlets. If a given lake had an original salt content, its age would have been overestimated. If proven correct, his idea was evidence that Earth was older than the ecclesiastics estimated.

While much of Halley's radical thinking on magnetism would later fail to withstand scientific scrutiny, his ideas on comets would endure and he personally would remain devoted to the hypothesis on which he based his magnetic theory that Earth was comprised of shells. It was a premise that would help unlock the secrets of the aurora borealis—those luminous clouds that form on the horizon, usually only in the most northern latitudes, and seemingly extended almost to the zenith.

ON MARCH 6, 1716, dazzling streamers of colored lights danced across the night sky. Halley—along with most of northwestern Europe—watched this aurora, then an unexplained phenomenon. It was the first he'd ever witnessed. His eyes were trained on the atmosphere until 3 a.m., though it was so intense it was visible in the daylight. Above his London home the clouds exuded rays of yellow, red, and a shadowy green. The colors of the rays grew more vivid the farther north one traveled in England and Scotland. The rays, as he noted, were often portrayed in religious paintings as the glory emanating from God.

Contrary to Gilbert's belief that the magnetic and geographic poles coincided, Halley had observed that they were skewed: Earth's magnetic pole was displaced from its geographic pole. The auroral phenomenon followed the magnetic pole. He had reckoned there must be a connection between Earth's magnetic field and the aurora.

Halley had reasoned that the rays correlated with the field lines of a uniformly magnetized sphere. He reckoned that the auroras were most intense at the Earth's poles, where the field lines converged, and surmised that the circulation of matter in the Earth's magnetic field caused the rays. (Essentially, he was right. Halley's later observations would also shed light on other aspects of geomagnetism related to this puzzle. Later, Anders Celsius and his student Olaf Hjorter would also observe magnetic disturbances linked to the polar auroras or northern lights; in our time such events are associated with magnetic substorms. And in the past few decades the study of the solar wind or how charged particles move in the magnetic fields of celestial bodies, including Earth, has become a leading field of space research.)

In 1724, after at last obtaining his dream job as England's second astronomer royal, Halley finally felt secure enough to publish his contentious paper written 30 years earlier. It discussed the deluge and the likelihood of Earth arising from the fallout of a former world, destroyed by the impact of a comet that led to the creation of the present world. According to the minutes of a Royal Society meeting at which he first delivered his hypothesis 30 years earlier, the collision would generate a new axis, rotation period, and length of the day and year and "account for all these strange marine things [fossils] found on ye tops of hills and deep underground."

Halley's views, in fact, were not necessarily antithetical to those of the Church of England, according to current interpretations of his later paper. Halley believed the world was co-eternal with God and dependent on him. But most of his contemporaries did not appreciate the nuance. The only satisfactory view to them was that the world was eternal and independent of God. For example, in Newton disciple William Whiston's 1696 *New Theory of the Earth*, an early treatise based on Newton's *Principia*, he wrote that it was "now evident that gravity (the most mechanical affection of body) and which seems the most natural, depends entirely on the constant and efficacious and if you will the supernatural and miraculous influence of almighty God."

But contemporary scholars contend that Newton was very much in support of Halley's idea that the current Earth with "visible marks of ruin upon it" had evolved from a previous one. Moreover, Newton may have influenced Halley's writings on the internal structure of Earth and its mutable orientation in space in his discourses with Hooke.

Yet it was his ongoing dialogue with Hooke that yielded much of Halley's scientific legacy on such key astronomical and physical topics as Earth's age, stellar cycles, comet cosmogony, and terrestrial magnetism.

WHEN NEWTON DIED in 1727, Halley may have been among the Royal Society members who escorted the body to its final resting place inside Westminster Abbey under a white and gray marble monument.

It supports a heavily decorated sarcophagus with a full-body sculpture of the knighted genius and a gilded relief panel detailing some of his key achievements.

Halley died some 15 years later on January 14, 1742, rather peacefully after imbibing a glass of wine. Both men lived to approximately 85, far longer than most in their day. But it was not long enough for Halley to see whether his comet would reappear.

In his will, Halley requested to be buried next to his wife of 55 years in the churchyard of St. Margaret at Lee near Greenwich and not far from the Royal Observatory. Mary Tooke had predeceased him by six years. Shortly after her death, Halley endured a minor stroke that left his right hand partially paralyzed and contributed to his gradual physical decline, but his mind was sharp to the end. His only surviving son, Edmond the naval surgeon, born the year the *Paramore* set sail, also died before him—but by only one year.

In his own lifetime, Halley's comet work was known by few. It held no more significance publicly than his papers on Palmyra or diving bells or the transit of Venus, for that matter. The comet was barely mentioned in his obituaries, which duly touted his work on enhancing navigation.

All that would change. Through 1758, as anticipation of the predicted comet's return grew, such astronomers as J. P. Loys de Cheseaux and even a Barbados plantation owner named Thomas Stevenson speculated that the irregularities in the periods were large enough that in fact two comets might be involved, each returning every 151 years. Still others suggested the period was decreasing arithmetically.

Interest in the comet's pending visit reached the New World. A 12-part series written by an anonymous author appeared in the *Boston Gazette*, a leading paper in Britain's colony of Massachusetts:

> That a comet is expected about the year 1758, is a thing that has been so often mentioned, and is indeed so generally known, that some may think it unnecessary to have any more said about it. But though it be known, it may not be generally attended to; and there is good reason to remind

people of it, because it is not at all improbable, that the comet may pass unseen, if it be not carefully watched for. . . . To prevent, if possible, its escaping Observation, which would be a very great disappointment to all lovers of astronomy, and to secure the earliest discovery of it that we can, we propose to publish in our paper, from month to month, an account of the apparent situation of the northern part of its orbit in the heavens, which is the only part in which it has ever yet been seen.

Before a mixed chorus of naysayers and believers in the public and the scientific community at large, the comet was sighted on Christmas Day 1758—16 years after Halley's death. A well-to-do German farmer and amateur astronomer, Johann George Palitzch, documented its return above a little village near Dresden. Meanwhile in Paris, astronomer Charles Messier independently sighted the comet on January 21, 1759.

Halley's forecast was dead on. The comet was not only delayed in its return due to the gravitational pull of Jupiter and Saturn, but it also appeared in the part of the sky Halley had foretold some 55 years in advance. It was not long before the gleam returning above Earth would be referred to as "Halley's comet" throughout Europe and eventually the rest of the world. Posterity would indeed not forget that the discovery was owed to an Englishman. For science, with regard to the public, the "second coming" of the comet would hardly be underestimated.

And at least for those who could understand the significance of its return, sighting a comet in the heavens would no longer be misinterpreted as an ominous sign of God's impending vengeance. (Such notions were instead replaced by a palpable fear that the physical fallout from a cometary impact could wreak havoc on the planet.) Moreover, the return of Halley's comet also affirmed Newtonian gravitational dynamics.

As Benjamin Martin wrote soon after the event in May 1759, "As it is the first comet which has been predicted, and has returned exactly according to that prediction, it cannot but excite the attention and admiration of the curious in general, and fill the minds of all astronomers with a ravishing satisfaction, as it has, by this return, confirmed Sir Isaac Newton's rational of the solar system, verified by the cometarian theory of Dr. Halley and is the first instance of astronomy brought to perfection." The intervals between the subsequent appearances of Halley's comet have ranged from under 75 to over 79 years, analyses reveal. Questions remain today as to all the forces, besides the pull of Jupiter and Saturn, that may affect the timing of its return near Earth.

In truth it turns out that only one of the 24 comets analyzed by Halley and published in the 1705 table was periodic. Later research would also reveal, for example, that Halley was wrong about his calculations concerning the comet of 1680. Its period was not 575 years. It likely won't return to tour the planet from the heavens any time soon. Its orbit is parabolic—and on the order of thousands of years.

Yet one correct prediction was proof enough.

Without Newton's work that he helped shape, Halley could not have forecast the periodic return of his comet or made many of his other key contributions to geophysics and astronomy. Halley would be among the first to apply Newton's principle of universal gravitation and resulting laws to the physical realm. Perhaps a lot of that had to do with the reality that many natural philosophers of Halley's day didn't immediately buy into Newton's theory since it entailed such phenomena as unexplained action at a distance.

Ironically, that comet of 1682 was the closest major comet preceding the *Principia* and would become a blazing symbol that forever connects the comet not only with the book that cracked its puzzle but also the seminal cooperation of Halley and Newton.

CHAPTER 13

LEGACY:
MORE THAN A COMET MAN

Halley probably never imagined the world his comet would return to when it reappeared in 1985. Astrophysicists from an American-based national science agency launched spacecraft to analyze it. Earthbound opportunists peddled cheesy T-shirts and other wares commercializing the event as much as commemorating it. Fringe groups and eccentrics circulated pamphlets warning of doom and gloom. Assorted reporters gave live accounts transmitted by satellite-feed or radio waves in every language of the comet's plodding approach to its perihelion. "Halleymania" imploded worldwide as February 9, 1986, the day when the celestial object would be closest to Earth, drew nearer and nearer.

To get the best glimpse of the spiraling ball of fire, a cottage industry of Halley-watching tours cropped up, offering a goofy gamut of "once-in-a-lifetime" opportunities. You could join science fiction writer Isaac Asimov and Czechoslovakian comet hunter Lubos Kohoutek aboard the luxury cruiser the *Queen Elizabeth 2* to listen to their lectures as you gawked at the comet from the North Atlantic,

dark skies and stormy seas included. Or fly above the clouds in a posh
charter jet to supposedly get an even closer view of the comet's flight.
Or "see the splendor of Halley's comet from the roof of Australia,"
according to one travel brochure. "Before you know it, the skies will
come alive with the most dazzling display of celestial bodies in 76
years," boasted another ad, promoting expeditions to South America
and South Africa, among other places.

The comet craze surrounding this luminous mass of ice and dust
was nothing new. Modern technologies and ambitions simply ampli-
fied the brouhaha. When the entity had revisited Earth's heavens in
1910, it was met with similar enthusiasm, just on a smaller scale. Instead of glow-in-the-dark T-shirts and doomsday cult memberships,
the spectacle was used to promote such items as fountain pens and
corsets. And the doomsayers were much less vocal and prolific. Despite the hoopla of the mid-1980s—the Halleywares market with its
several thousand product offerings was cautiously pinned at $500
million—the conditions for watching the comet from Earth were the
poorest they'd been in two millennia, and it made a disappointingly
faint and unspectacular showing.

Today, Halley's comet—transformed by the first triumph of the
Newtonian revolution from a dire supernatural omen to a predict-
able element of the universe's clockwork—remains a recurring symbol of this scientific age of the Enlightenment. His comet is hurtling
through space at some 20,000 miles per hour and won't be back until
2061. But it can remind us of past epochs and everlastingly of Halley's
contributions to geophysics and the world of science writ large.

Halley was a man of diverse talents. As Royal Society scientist
and administrator, Oxford professor, and eventually England's Astronomer Royal, his influence, like his perspective, was wide, touch-
ing many reaches of society. In his lifetime, Halley published 80
far-ranging contributions in the prestigious *Philosophical Transactions*. As one of his country's leading minds of the Enlightenment
and its related revolutions, he changed the course of science, industry, and related policy making. In many ways he was an artist of the

transformation that was under way, reflecting, defining, and shaping his age. A generalist in the purest sense, this sage sea captain often mistaken for a pirate was at once an adventurer and a quiet observer, brilliant navigator and pioneering actuary, justice seeker and secret warrior, meticulous historian and forward thinker, avid inventor and methodical referee, skilled cogitator and civic servant, fierce rival to several of the nation's privileged brainiacs and close confidant to other greats, including Sir Isaac Newton.

As Halley biographer MacPike put it: "The idea of commissioning a landsman to the command of a King's ship might appear to professional seamen as sheer madness. . . . However, in Halley's case the rash outrageous act was justified in the event, but only because Edmond Halley was one man in a thousand, possessed of the most varied gifts and the most extraordinary versatility."

With Halley's assistance on ushering the publication of *Principia* in 1687, Newton altered the way the world perceived the universe. In fact, to many scholars this scientific giant embodied the dawning Enlightenment, broadly identified from the "Glorious Revolution" of 1688 that installed William and Mary as monarchs of England to the French Revolution a century later.

By the time Halley embarked on his Atlantic Ocean expedition, Newton's discoveries about gravity and light had unlocked God's laws of the universe for those in the know. At the same time, the heated Battle of the Books was already raging in smoke-filled coffeehouses, in cloistered academic chambers, and on rudimentary printing presses. Literary scholarship with its recognized historical underpinnings was pitted against natural philosophy with its modern ideas. Diplomatic, if not debonair, Halley straddled both cultures. Schooled in the classics yet chasing scientific ideals, he was interested in tinkering in all worlds of discourse.

Revering knowledge, education, and opportunity, Enlightenment crusaders of Halley's ilk instilled new values. Truth, freedom, liberty, and progress were the buzzwords of the day. Disciples of Locke and his greatest patron, the Earl of Shaftesbury, spread the word about civic responsibility. They believed in an innate duty to act publicly for the greater good of humankind and the rising British nation. Like many of the movement's movers and shakers, Halley had taken this decidedly British notion to heart with exceptional ardor.

England's growing prosperity and global reach would foster a social climate that would advance the ideals of science and its application to commerce and industry. The restoration of King Charles II served to bring the scientific movement "into the open daylight of fashion and favor." He supported the founding of the Royal Society, and it was during his reign that Halley matured intellectually and came to be intimately tied to this erudite club.

As time wore on, the increasing openness of the island nation's religious climate would enable the promotion of observation, experimentation, and reasoning to achieve science-based findings about the natural world. It was out of this environment that Halley secured his mission and in which his career soared.

Clearly, Charles's successors, Queen Mary and King William, influenced by the elite minds of the Royal Society, believed that a "mission" of science could incite progress. That was an amazing leap of faith for a monarchy. And much was at stake. Any failure of Halley's international mission of magnetism could have irreparably shaken confidence in public ventures in science, possibly forever dampening Parliament's willingness to fund bigger and better endeavors in the future.

A spirit of inquiry propelled Enlightenment players to "discover the world." Halley personified this essence in practice. Though often in Newton's shadow, Halley never outwardly showed any signs of jealousy. His enthusiasm for his colleagues and the advancement of science went unchecked. Regardless of any wooly politics that were afoot, Halley was morally compelled to seek and share knowledge in order to further humankind and with it the position and prestige of the British nation. With his broad and at times even divergent perspective and willingness to undertake personal risk, he fulfilled a role that neither the reclusive and erratic Newton nor their other cantankerous contemporaries could play.

If the so-called Scientific Revolution indeed began in 1600 with the publication of William Gilbert's *De Magnete*, from which Halley took a magnetic cue, and finished in 1714 with the passing of the Longitude Act, Halley's interests and career bridged the movement. In England, in particular, the revolution's impetus was nautical—and based on practical needs. The Royal Observatory, where Halley started as a precocious young investigator and ended his career as the second royal astronomer, was expressly founded "for perfecting the art of navigation."

The government, with Halley's prompting, was promoting not faith but reason based on science. Or at least Parliament was sanctioning a new sort of faith, not in religion but in the method of experiment and observation, on which the Royal Society itself was based, when it offered a huge reward for the person who could solve the longitude problem. The challenge would intrigue Halley all his intellectual life. Yet few projects would capture the essence of what the Enlightenment was all about better than his mission aboard the *Paramore.*

The rise of national trading companies spurred on the revolution in science and the Enlightenment. Just as Halley had fervently pushed for advances in navigation, emerging English trading companies had helped persuade the monarchy that control of the seas was paramount. The Elizabethan mariner Sir Walter Raleigh himself had said it well: "Whosoever commands the sea, commands the trade; whosoever commands the trade of the world, commands the riches of the world, and consequently the world itself."

THE PROLIFERATION OF BRITISH TRADING companies also enabled the nation to control foreign lands in exchange for its support of such enterprises. As a result, England's influence on local social, political, and economic matters in its colonies and territories only grew. Naval and

military support usually followed the creation of such diplomatic al-
legiances. This in effect passed the task to the British Navy of molding
the world to the monarchy's desires. The responsibility for ensuring
free trade and access to markets, which are still the key elements of
globalization today, fell to these water warriors for whom Halley
strove for much of his life to give a competitive edge by advancing
navigational knowledge and tools on as many fronts as possible.

The use of Enlightenment ideas to rationalize such intervention
on foreign soils—no matter how well intentioned—illustrates the inherent contradictions of thought and action throughout the movement that had cross-cut all strata of British society. For example, the
spirit that brought the revolution of 1688 and William and Mary to
power was soon compromised by internal political realities and power
struggles between the more conservative Tories and the Whigs who
sought reform. Whether the changeover of power rescued England
from the throngs of tyranny—or quickly enough—remains an open
question. For the most part, all citizens remained subjects with effectively few basic rights as such, at least by modern standards.

Halley himself embodied personal and public contradictions. His
classist commentary on life expectancy, for one, was telling. Perhaps,
British society wasn't prepared for a scholar who swore or a sailor
who wore a periwig, a truth seeker who wagered his life for a few data
points, or a man faithful to God but who put faith in scientific methods before the tenets of established religion. In the end, whether
through his dogged persistence or at times sycophantic pandering to
patrons, all society would embrace him.

Just the same, these new ways of thinking and behaving entailed
by the Enlightenment touched all aspects of day-to-day life, inter-
leaving history, art, science, philosophy, politics, and religion into the
tapestry. In this period the culture of the educated person came to
embrace the whole rubric of human knowledge. It was an age of the
polymath when science and philosophy were not distinct from theol-
ogy. And generalists thrived. To be educated was to be an artist, scien-

tist, historian, and philosopher in one. And Halley was among the Enlightenment's champions.

THROUGHOUT HIS LIFE, HALLEY HAD the patronage and even the admiration of a long line of kings and queens and others from the nation's influential classes—and the respect of much of the leadership of the Western world. The importance of Halley's personality to the promotion of his career cannot be underestimated. Few others managed to secure the royal patronage of seven monarchies spanning nearly 70 years of swirling political tumult.

Unlike Halley, Samuel Pepys, for one, fell out of royal grace after James II abdicated the throne. As secretary of the Admiralty during many of the British monarchy's turbulent years leading up to James's departure, he kept the British navy a league ahead of rival nations by streamlining its bureaucracy and advocating for the necessary resources to keep its ships better maintained and outfitted. But his career and influence suffered when William and Mary came to power since he'd curried only James's favor. Pepys, though, was already in his mid-60s when the *Paramore* set sail.

Halley's ability to charm the powerful of many persuasions furthered his ambitions in other ways. While most astronomers to his time were reputed solitary men, Halley maintained a following of friends, colleagues, and acquaintances. A leading scholar dubbed him "a man of prodigious versatility and most attractive personality." The 1757 *Biographia Britannica* credits him with "the qualifications necessary to obtain him the love of his equals. In the first place, he loved them; naturally of an ardent and glowing temper, he appeared animated in their presence with a generous warmth which the pleasure alone of seeing them seemed to inspire." The distinction set him apart from his more unsociable peers like Flamsteed and Newton.

Halley used his diplomacy for the sake of learning. He encouraged the circulation of data and findings and intellectual discourse to stimulate new lines of investigation.

IN HIS AT TIMES SELF-EFFACING ROLE as clerk to the Royal Society during his early career, Halley was the organization's most prolific administrator in terms of published papers, and his unpublished ones were often more remarkable. He was his age's most dedicated servant to knowledge and perhaps to the public. He selflessly put the work of other more senior members before his own. He helped perpetuate the society itself by settling disputes among its members and keeping the presses rolling even in hard times. He was a de facto science advisor to the Royal Navy and the monarchy as a hydrographer, meteorologist, mint official, and more. He was also a diplomat and ambassador—and not only of the scientific realm.

More than a mere theoretician, Halley excelled at applying science and mathematics as well. Among his other coups, his practical gifts to navigation, at bare minimum, are worth mention. He designed a new reflecting gadget to measure altitudes to obsolete the Davis quadrant. He invented a glass point to better prop up a ship's compass card. He also improved the log line by developing an instrument to measure the speed of a ship based on the angle of inclination of a line tossed from a ship's stern and attached to a brass bell.

He also devised a method to compute meridional parts to make Mercator charts, among other contributions. As astronomer royal, Halley, like his predecessor Flamsteed, can be partially credited for helping spawn the business of scientific instruments and chronometry in England and the resulting improvements in observation and measurement.

MANY OF HALLEY'S MORE ABSTRACT navigational dreams would be fulfilled after his death, but all would bear his fingerprints. Some 70 years after Halley's voyages, Captain James Cook would bring three key aspirations of Halley's career full circle. Cook's first voyage was spawned by Halley's original prediction of the transit of Venus. Cook would use the lunar method for finding longitude that Halley had chased his entire intellectual life. On his second and third voyages, Cook would also be abetted by a device for determining longitude, the Kendal watch, which was a direct outgrowth of the Harrison clock that Halley had supported when few others, Newton among them, kept an open mind. (Based on Halley's advice, top instrument maker George Graham lent Harrison enough money to allow him to make his first prototype to submit to the Board of Longitude.)

When Halley was sailing his *Paramore* across the North Atlantic and making the first charts of geomagnetism, little did he ever imagine magnetism would underpin today's stunning advances in information technology and electromagnetic engineering. Magnetism also offers ways to study phase transitions, random disorder, and physics in low dimensions, which looks at particle interactions at ever higher energies in order to essentially study matter at smaller and smaller size scales.

But perhaps just as astonishing was Halley's lack of progress on other fronts. Could Halley ever fathom that three centuries after his death many mysteries of Earth's magnetism would remain unsolved? In the decades leading up to the 21st century, study of the solar wind emerged as a leading field of space research. Many fascinating phenomena of magnetic fields remain to be explained in chemistry and biology as well.

Halley might relish the fact that geomagnetism still flummoxes legions of theoreticians and fact finders. While many of Halley's ideas about magnetism wouldn't pan out, he remained personally committed to his hypothesis that Earth was comprised of concentric shells. Today geophysicists know that Earth's magnetic field is generated by electric currents deep within the surface and high above it. They have long known that Earth's total geomagnetic field is the result of the superposition of all these currents; its model is much more complicated than Halley ever envisioned. The main driving force behind Earth's magnetic field comes from electric currents sustained by a dynamo in the core. The kinetic energy from the motion within the vast volume of molten iron fluid circulating in the core creates the electric and magnetic fields much like the spinning wire coils in an electric generator. Although scientists know the so-called geodynamo must be regenerative, they are still unraveling the mysteries of how it works.

While most seafarers take it for granted that their compasses point north, Earth's magnetic field has actually reversed itself hundreds of times in the past 4.5 billion years. But scientists are yet to understand what makes the field flip from north to south and back again on average every 250,000 years.

It's no wonder that Earth's magnetic variation caused many throbbing headaches for the world's mariners over the centuries. Its complexity accounts for why the amount of declination also varies depending on geographic location and time.

And we should remember how Halley pushed this line of inquiry in new directions. His comprehension of Earth's magnetic field, which he conveyed through the curved lines of his hard-won map, and interpretation of the meaning of the resulting global patterns of the field pointed research toward unexplored territories. With his "metamap," he successfully boosted notions of situational awareness while en route and laid the groundwork for today's constellations of global positioning systems. He realized that Earth's magnetic field likely originated deep inside the planet and that its magnetic field lines were correlated with auroral phenomena. He also correctly surmised that the westerly drift of the magnetic variation had no impact on Earth's external dynamics.

All of which remains relevant today.

HALLEY'S VISION WAS RIVALED only by his appreciation for what had gone before him. An adventurer at heart, he was passionate to the end of his life about chronicling and understanding change of all sorts and how it related to humankind. From questions about the cycles of the Moon and tides to variations in Earth's magnetism to the causes of ancient floods to the timing of Julius Caesar's landing, he considered them all. And more significantly, he was willing to risk his life, family, personal money, career status, and more to obtain answers. There were his excursions to the enchanting island of St. Helena, the icy reaches of the Antarctic Circle, the bottom waters off the West Sussex coast, and more.

Perhaps unintentionally, Halley's pursuits point us toward a reconciliation of sorts of the ongoing conflicts between the study of the empirical natural world and the study of human values. We live, it seems, in a world that at once limits human potential yet at the same time offers boundless opportunities.

In his own lifetime, few in the general populace knew of Halley's comet work. Today his comet often overshadows his other assorted contributions. "We do not often think of him as a sailor; and yet, previous to Cook, Captain E. Halley was our first scientific voyager," is how historian S. P. Oliver put it in 1880.

We often forget his scientific values and foresight as well as his intuitive perspectives on everything from auroral phenomena to the source of Earth's magnetism. Halley's perseverance in his wide-ranging global sea mission could not be more relevant to an age where cooperative worldly initiatives are hardly optional in research as well as other realms. International surveys and joint observations are increasingly mandated by societal needs and objectives.

Yet Halley's international legacy endures thanks in large part to that seeming fireball in the sky. Once every 76 years, more or less, the whole world looks up, beyond the horizon and seems unified in its attention to the heavens for a few fleeting moments, focused on something that still startles our imagination. At a minimum, even today we all wonder where the course of science and human progress, which Halley forever changed, will have arrived in 2061 when Halley's comet next returns.

Maybe by then the Battle of the Books will finally be resolved. And conventional wisdom might hold that Halley was much more than a paid stargazer who had a comet he didn't even discover named for him.

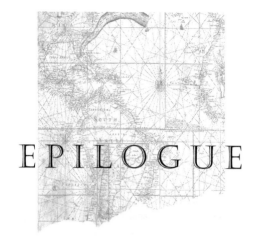

EPILOGUE

At the Royal Observatory on a bright, sunny day in 2004, gaggles of tourists, drawn from London by riverboat ride on the Thames from the Millennium Bridge or by rail through the docklands, straddle the prime meridian. The world's starting and finish line seems to have an almost magnetic attraction for them. To this student of Halley's adventures, the longitude marker conjures a mixture of both his failed and successful quests. It was his quest to solve the problem of longitude that drew him to the Royal Observatory during his undergraduate days—and that quest on which he would write one of his final papers and continue working to the end of his life. He used an array of methods to calculate it; some were more dependable than others. His lunar approach, of course, proved impractical in the long haul. While his work may have come close to providing a workable method in his lifetime, he never personally staked a claim to the longitude prize.

The great thing about science is that negative discoveries can be transformative, and his enhancement of the world grid's magnetic accuracy was a critical stage in an ongoing process. Indeed, his quest for knowledge through wide and open collaboration is a model for today. In time his other contributions to the world of science would overshadow much of his purely astronomical work, though it would be the comet that would immortalize his name.

Here, I see the Latin inscription on the capstone of his tomb, which was moved to Greenwich in 1845 and inlaid in a wall of a building connected to the actual observatory. It reads in translation:

> Under this marble peacefully rests, with his beloved wife, Edmond Halley, L.L.D. unquestionably the greatest astronomer of his age. But to conceive an adequate knowledge of the excellencies of this great man, the reader must have recourse to his writings, in which all the sciences are in the most beautiful and perspicacious manner illustrated and improved. As when living he was so highly esteemed by his countrymen, gratitude requires that his memory should be respected for posterity. To the memory of the best of parents their affectionate daughters have erected this monument in the year 1742.

Halley's corpse still rests below a now-obscure slab in a cemetery in what now belongs to a London suburb. His grave outside the parish church of Lee is unkempt. Although the capstone of his tomb is now prominently displayed at the Greenwich Observatory, some view the initial stature of Halley's grave as one last rebuff by the Church of England.

Halley has since been honored in myriad ways. A crater was named for him on the Moon, at 8 degrees south and 6 degrees west, as was the Royal Society's permanent scientific base in Antarctica. It was named Halley Bay in 1957. A bronze memorial in the shape of a comet was erected at Westminster Abbey, London's Who's Who of shrines, on November 13, 1986, to culminate celebrations of his comet's most recent return.

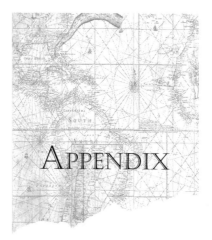

APPENDIX

The following appendix comprises a random sampling of surviving documents related to Halley's three voyages. They include journal entries from his voyages, correspondence between Halley and the Secretary of the Navy Burchett, lords letters to the Navy Board, treasury papers, and Navy Board minutes, as well as the text of his description that accompanied some publications of his first sea chart.

Royal Society Council Minutes
Vol. 2, p. 146
15 February 1698/9

It was proposed and balloted that Mr. Halley being absent upon a particular Service and no provision being yet made by the Council for Supplying his place; It was Ordered that Dr Sloan Should propose the Names of two or three persons to be Assistant to the Secretaries as Clerk; And lit was carried in the Affirmative Nemine Contradicente.

It was Ordered in the mean time that Dr. Sloan Should employ whom he pleases any where else but none at the Table—Where the Society meets but whom the Council Agree to.

From Halley's journal of the first voyage:

February 17, 1698

This Morning between two and three looking out I found that my Boatswain who had the Watch, Steard a way Nw instead of W (we no baring down W for the Iseland of Fernando Loronho) I conclude with a deigned to miss the Iseland, and frustrate my Voyage, though they pretended the Candle was out in the Bittacle, and they could not light it. About 3. In the afternoone we made the Iseland about 6 Leagues off us; bearing SWBW: The next day we came to an Anchor under the Lee of the Island, having narrowly escaped a Sunk Rock; that lies off the Sw point of the Island. I went on Shore to see what the Iseland might afford us, but found nothing but Small Turtle Doves and Land Crabbs in abundance, neither Goats nor hogs nor any people; we saw many green Turtle in the Sea and in Someplaces their Tracks on the Sand, but could Catch none, by reason of the great Surf of the Sea; we searcht the whole Lee Side of the Iseland but found no fresh water; we lookt not on the windward side because we found such a Suff on the Lee side; here we againe scrubb'd our Shipp and gott some Wood and Sett up all our Shrouds and brought our masts more aft. We found a four Clock Moon to make high Water, and it flows about 6 Foot on a Spring. The Variation observed on Shore was not full 3 degrees East. The Island is but Small, about 7 Miles Long and very Narrow. The Middle thereof is in the Latt 3 degrees 57' South, and Longit by reckoning from London 23 degrees 40' West. The Appearnace thereof when the high pico like a Steeple bears SWbW at 5 Leagues distances is thus————————

Letter from Halley to Burchett:

Captain's Letter Book
To the Honourable Josiah Burchett Esquire
Secretary to the Admiralty of England
These – humbly present London

Honoured Sr.

I have had no opportunity to give their Lordships any account of my proceedings since my last of December 20 from Madera. That same day I sayled for the Cape de Virde Islands and arriving at St. Jago on January 2, I found there two English Marchat shipps, one of which called the New Exchange, whereof one John Way is Master belonging to London, was pleased, instead of saluting us, to fire at us severall both great and small shott. We were surprised at it, and beliving them to be pirates, I went in to windward of them and bracing our head sailes to the Mast, sent my boat to learn the reason of their firing. They answered that they apprehended we were a pirate, and that they had on board them two Masters of vessels, that had been lately taken by pirates, one of which swore that ours was the very ship that took him; whereupon they thought themselves obliged to do what they did in their own defence. Then they sent on board me the two persons they said were the Masters of the taken Vessells, and

soon after the two Masters came themselves, they said they were sorry
that they had fired at the Kings Coulours, but that colours were not
to be trusted. I told them I must acquaint their Lordships with what
had past, and if their Lordships would put it up, as it hapned they had
done me no damage. The next morning they both sailed, and upon
our arivall here we found the said Master John Way and his ship in
the road. From St. Jago we proceeded to the southward and being
gotten within 100 leagues of the line, we fell into such calmes and
small southerly gales, that our ship being very indifferent to windward, we were full seven weeks before we gott 100 leagues to the
Southward of the line, in which time our water being near spent,
obliged us to recruite it on the coast of Brasile. By this time twas
March and we found the Northerly Currents made against us, and we
upon the Lee-shore; so that it would have been scarce possible for a
more winderly ship than we, to turn it to the Southward. And the
winter advancing apace in those Climates I principally entended to
discover, I thought it not adviseable to proceed that way at this time
of the year; hoping it may give their Lordships some satisfaction if I
do curiously adjust the Longitude of most of the Plantations and see
what may be discovered in relation to the Variation of the Needle in
the Northern Hemisphere. Twas the last of November before we left
the coast of England, wch considering the uncertainty of the Winds
was I find above two months too late; but I hope to be in England
time enough to proceed again this year if their Lordships shall think
fitting to allow it. We watred in the river of Paraiba in Brasile here the
Governour Dom Manuel Soarez Albergaria was very obliging and
civill, but the Portuguez, as farr as I could guess, were very willing to
find pretences to seize us and tempted us severall times to meddle
with a sort of wood they call Poo de Brasile which is an excellent dye,
but prohibited to all foreigners under pain of confiscation of Shipp
and goods. I being aware of their design absolutely refused all commerce with them, and having gotten our water we arrived here in
three weeks, on the second of this month; Our whole shipps company is hither in perfect health and our provision proves very good.

I am

Honoured Sr

Your most obedient Servant

Edm. Halley

Paramour Pink
In Barbadoes road
4 April 1699

Received 9 June in ye morning mail

Letter from Halley to Burchett:

Captain's Letter Book
The Honourable Josiah Burchett Esquire
Secretary to the Admiralty of England
These present London

Honoured Sr,

I this day arrived here with his Majesties Pink, the Paramore in 6 weeks from the West Indies, having buried no man during the whole Voiage, and the Shipp being in very good condition. I doubt not but their Lordships will be surprised at my so speedy return, but I hope my reasons for it will be to their satisfaction. For as, this time, it was too late in the year for me to go far to the Southwards, I feared that if I went down to Jamaica, and so to Virginia &c. the same inconvenience of being late might attend me in case their Lordships, as I humbly hope, do please that I proceed again for I find it will be absolutely necessary for me to be clear of the Channell by the end of August or at farthest by the middle of September. But a further Lieutenant, who, because perhaps I have not the whole Sea Directory so perfect as he, has for a long time made it his business to represent me, to the whole Shipps company, as a person wholly unqualified for the command their Lordships have given me, and declaring that he was sent on board here because their Lordships knew my insufficiency. Your Honour knows that my dislike of my Warrant Officers mad me Petition their Lordships that my mate might have the Commission of Lieutenant, therby the better to keep them in obedience, but with a quite contrary effect it has only served to animate him to attempt upon my Authority, and in order therto to side with the said officers against me. On the fifth of this month he was pleased so grosly to affront me, as to tell me before my Officers and Seamen on Deck, and afterwards owned it under his hand, that I was not only uncapable to take charge of the Pink, but even of a Longboat; upon which I desired him to keep his Cabbin for that night, and for the future I would take charge of the Shipp myself, to shew him his mistake; and accordingly I have watcht in his steed ever since, and brought the Shipp well home from near the banks of Newfound Land, without the least assistance from him. The many abuses of this nature I have received from him, has very sensibly toucht me, and made my voyage very displeasing and uneasy to me, nor can I imagine the cause of it, having endeavoured all I could to oblige him, but in vain. I take it that he envys me my command and conveniences on bord, disdaining to be under one that has not served in the fleet as long as himself, but however it be I am sure their Lordships will think this intolerable usage, from one who ought to be as my right hand, and by his example my Warrant Officers have not used me much Better; so that if I may hope to proceed again I must entreat their Lordships to give me others in their room.

Notwithstanding that I have defeated in my main design of discovery, yet I have found out such circonstances in relation to the Variation of the Compass, and the method of observing the Longitude at Sea (which I have severall times practiced on board with good success) that I hope to present their Lordships with something on those articles worthy of their patronage. I humbly entreat yr Honour to expedite my orders in the Downs, and if it be their Lordships pleasure, that the Shipp continue there for some time, they please to give me leave to come up to waite upon them, to give them a fuller account.

I am

Their Lordshipps and

Your Honours most obedient servant

Edm. Halley.

Plimouth
23 June 1699.

Letter from Burchett to Halley
Secretary's letter book

Admiralty 29 June 1699.

Sr:

I have received your Letter of ye 23rd inst from Plymouth & this comes to meet You in ye Downes to acquaint you that Orders are sent to Sir Clo: Shovell to try your Lieutenant at a Court Martial upon ye Complaint made of him in your said Letter. To which I have only to add, that when ye matter is over You will receive Orders from Sir Clo: Shovell for repairing to Longreach and from thence to Deptford where she is to be paid off & laid up. I am

Yr &C.

J.B.

Capt. Halley—Paramour Pink—Downes

Wages book extract:

Sber	2	Jon Dunbar	Midspn	
	9	Tho: Price	Carp	
	10	Jam: Glenn	Ab	
		Jon Hughes	Carp S	D 16: Sber 98
	12	Fra: Thracia	Ord	D 19: D 98
		Davd Wishard	Ab	
		Richd Arnold	Ab	D 20: Sber 98
		Jam: Garret	Ord	D 25: Sber 98
		Jam: Canadie	Ord	R 13 Sber 98
		Tho Baley	Ord	D 1 Ober 98
	14	Edwd Child	Ord	D 25; Sber 98
	19	Sam: Withers	Ab	
	21	Wm Dowty	CarpMat	
	22	Geo: Alfrey	Chyr	
	26	Wm Harrison	Ab	D 14: Ober 98
		Cleb Harmon	CaptCl	
Sber	26 98	Wm Edwards	Ord	R9 Ober 98
	29	Wm Jones	Ord	R7 Ober 98
		Tho; Daviss	Ab	
Ober	10 98	Edwd Harrison	Mat & Lieu	
		Tho: Burton	St	
	17	Robt Dampster	Carp St	D 24 June 99
	19	Rd Pinfold	Capt St	
		Dan: Dewett	Ab	D 22: Sber 98
Ober	20	Jon Vinicot	Ab	D24 June 99
Mar 8 1698/9		Sam Robinson		D3: Apr 99
Apr 16 99		Hen: Clarke		Ab
July 9 99		Tho: Paramour		Carpt St

[signed]
Edmond Halley
Edw: Harrison
George Alfrey

Nett Book

Paramour Pinck
Began Rigg: Wages   15th August 98
Sea ....D   31 October 98
Ended Wages   29 July 1699
Being then paid off at Broadstreet
Present
    Sir Richard Hddock Kt
        John Clarke   Clerk
        Wm Hogg   Clerk
Read the 21 November 1700
Made up 23 December
    Full————————————   520.   2.   3.
    Deductions—————————   43.   16.   10
Neat sume paid to ye 23d Novem 1700   476   5   5

The Rt Honourable Edward,
Earle of Orford Treasurer

Halley's journal entry for February 1, 1700, from his second voyage:

Latitude by Account 52 degrees: 24 minutes. Yesterday in the Afternoon with a fresh Gale at N b W, I steard away E S E, and between 4 and 5 we were fair by three Islands as they then appeard; being all flat on the Top, and covered with Snow. Milk, white, with perpendicular Cliffs all round them, they had this appearance, and bearing.

The greate hight of them made us conclude them land, but there was no appearance of any tree or green thing on them, but the Cliffs as well as the topps were very white, our people called A by the Name of Beachy head, which it resembled in form and colour, and the Island B in all respects was very like the land of the North-foreland in Kent, and was as least as hight and not less than five Miles in Front, The Cliffs, of it were full of Blackish Streaks which seemed like a fleet of Shipps Standing out to us. Wind blowing fresh, and night in hand, and because our vessell is very leewardly, I feard to engage with the Land or Ice that night, and having Steard in as farr as I durst, I resolved to Stand off and on till day, when weather permitting I would send my boat to See what it was. In the night it proved foggy, and continued so till this day at noon, when by a clear glare of Scarce _ of an hour we saw the Island wee called beachy head very distinctly to be nothing else but one body of Ice of an incredible hight, whereupone we went about Shipp and Stood to the Northward. True Course to this day noon is S 44 E. 25 Miles. Difference of Longitude 29 Minutes East: Longitude from London 35 degrees: 13′.

Letter from Halley to Sir Robert Southwell, friend of Samuel Pepys, on how to survey a coast:

27 January 1702 London

[Address:]   from Captain Halley
             to Sr R. Southwell;
             About taking the Survey of a Coast

Honourd Sr.

In obedience to your Commands I have endeavored to draw up such plain directions for making the Survey of a Coast, as may be serviceable to any that have the will and opportunity to describe curiously any Shoals they are acquainted with;

In order to this Survey of a Sea coast and to lay down truly the shoals and dangers near it, if the land be accessible the best way will be to take with all possible care the true positions of as many remarkable objects such as Steeples, Mills, Rocks, Cliffs, Promontorys, or such like as you find most conspicuous along the coast, that is their true barings from one another in respect of the true North and South; which is best done by measuring the angle with any proper instrumt.

From the rising or setting sun, allowing his amplitude and according to the exactness of these angles will your survey be more or less true. I prefer this method of taking these angles by the Sunn rather than by the compass or magneticall needle, because of the smalness of the radius of the Magneticall chart and the uncertainty of the variations on the Lands, the needle being affected with the neighbourhoods of Iron Oars and Mineralls.

This done you may readily plot down all those objects on the Land, by any view of them from a vessell riding at Anchor off at Sea; for if you take their true position from your ship, by help of the rising or setting Sunn as before, the intersections of those lines with those of the positions of the objects to one another, will give you the places and proportionall distances of the sd Objects one from another, to which afterwards a scale may be adapted, as shall be taught by and by.

Being thus assured of the plot of several objects on the shore, it will be very easy to lay down the points of any sand or shole, or any sunk rock on that Coast, either by the position of two or more of those objects, from a vessell riding at those points; or more compendiously and easily by taking the angles between those objects, at the said places entended to be laid down in your platt. That this may be the better understood, take the following Scheme. On the Coast to be described Let A be a steeple, B a Mill, C a Rock, D a remarkable Tree, E a steep Cliff&c. and lett each of them be seen from some other of the objects, and their positions truly taken. For example Let B bear from a West 12 degrees. Northerly; C from B, W 30 degrees. Southerly, D from C W. 20 degrees. N. and E from A W 2 degrees. Southerly. This done at a convenient distance off at Sea as at [the point marked by the arrow, let the position of A be North 20 degrees Easterly, B. N 2 degrees East, C. N 22 degrees $^1/_2$ W or NNW, D. N 40 degrees W and E. N 53 West. I say the true platt of the aforesd places A, B, C, D, E &c. will be as in the Scheme, although as yet we know not the reall distances of them, and wee may use them with certainty to lay down any other places in their true position. As for example, let there be a shole at one end whereof E bares NNW and C, NE; this being protracted, tis plain tis at F that that point of the shole ought to be laid down. At the other end of the same shole which wee will call G, for want of the Sunn, wee can only take the angle DGB 60 degrees & the angle CGA 80 degrees, I say the point G will be nicely determined therby. For if the angle BDH be made 30 degrees or the Complement of DGB, DH=BH, the arch of the Circle DBG described within the radius DH and center H, shall be such that wherever you take the point G therin the angle BGD shall be 60 degrees. In like manner making the Angle ACK 10 degrees the arck of the Circle AGC whose center is K passing through A and C, shall in all its points G make the angle AGC 80 degrees, and consequently the Intersection of those two Arches is the point G sought; this is demonstrated from 20.3. Euclid: the angle at the Circonference of a Circle being half of that at the Center. This is a very easy and expeditious way for putting down the soundings in Sea Charts in their proper places, and may be practisd in a ship under saile.

If it be an enemys Coast or otherwise inaccessible, it wil be necessary to make use of two Shipps or Boats, as two Stations, wherby to obtain the position of the objects on shore; which afterwards you may use as before. After your chart is made, you may adapt a scale to it, by help of the motion of Sound, which has been accurately tried both in England and France, and it is certain that sounds be they great or small move at the motion of Sound, which has been accurately tried both in England and France, and it is certain that sounds be they great or small move at the rate of a marine League in 15 seconds of time; and in still weather a gunn may be heard a great way, especially before a gentle gale of wind, and this I propose and recommend as a very usefull method of determining distances in these Hydrographicall Surveys. I shall be very willing further to explain any thing herin, that may obscure or difficult.

I am Your Honrs. Most obedt. Servt.
Edm. Halley

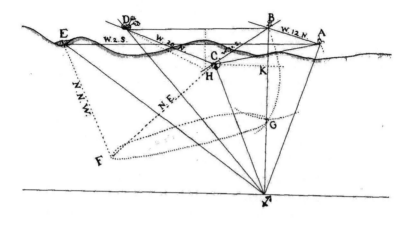

Halley's description accompanying his Atlantic Chart.

The Description and Uses
Of a New and Correct
SEA-CHART
Of the Western and Southern
OCEAN
Showing the Variations of the
COMPASS

The projection of this Chart is what is commonly called Mercator's; but from its particular Use in Navigation, ought rather to be named the Nautical; as being the only true and sufficient C H A R T for the Sea. It is supposed, that all such as take Charge of Ships in long Voyages, are so far acquainted with its Use, as not to need any Directions here. I shall only take the Liberty to assure the Reader, that having taken all possible Care, as well from Astronomical Observations, as Journals, to ascertain the Scituatuion and Form of this Chart, as to its principal Parts, and the Dimensions of the several Ocean; he is not to expect that we should descend to all the Particularities necessary for the coaster, our Scale not permitting it

What is here properly New, is the Curve-Lines drawn over the several Seas, to shew the Degrees of the Variation of the magnetical needle, or Sea Compass; which are design'd according to what I my self found in the Western and Southern Oceans, in a Voyage I purposely made a the Publick Charge in the Year of our Lord 1700.

That this may be the better understood, the curious mariner is desired to observe, that in this Chart the Double Line passing near Bermudas, the Cape Verde Isles, and Saint Helena every where divides the East and West Variation in this Ocean, and that on the whole Coast of Europe and Africa the Variation is Westerly, as on the more Northerly Coasts of America, but on the more Southerly Parts of America 'tis Easterly. The Degrees of Variation, or how much the Compass declines from the true North on either side is reckoned by the number of the Lines on each side of the double Curve, which I cal the Line of No Variation; on each fifth and tenth is distinguished in its Stroak, and numbered accordingly, so that in what Place soever your Ship is, you find the Variation by Inspection.

That this may be the fuller understood, take these Examples. At Madera the Variation is 3 and $1/2$ d. West; at Barbadoes 5 $1/2$ d. East; at Annabon 7d. West; at Cape Race in Newfoundland 14d. West; at the Mouth of Rio de Plata 18d. East, &c. And this may suffice by way of Description.

As to the Uses of this Chart, they will easily be understood, especially by such as are acquainted with the Azimuth Compass, to be, to correct the Course of Ships at Sea; For if the Variation of the Compass be not allowed, all Reckonings must be so far erroneous; And in continued Cloudy Weather, or where the mariner is not provided to observe this Variation duly, the Chart will readily shew him what Allowances he must make for the Dfault of his Compass, and thereby rectify his Journal.

But this Correction of the Course is in no case so necessary as in running down on a Parallel East or West to hit a Port; For if being in you latitude at the Distance of 70 or 80 Leagues, you allow tot the Variation, but steer East or West by Compass, you shall fall to the

Northwards or Southwards of your Port on each 19 Leagues of Distance, one Mile for each Degree of Variation, which may produce very dangerous Errors, where the Variation is considerable; for Instance, having a good Observation in latitude 49d, 40m. about 80 Leag. Without Scilly, and not considering that there is 8 Degrees West Variation, I steer away East by Compass for the Channel; but making my way truly E. S.d N. when I come up with Scilly, instead of being 3 or 4 Leagues to the South thereof, I shall find my self as much to the Northward: And this Evil will be more or less according to the Distance you sail in the Parallel. The Rule to apply it is, That to keep your Parallel truly, you go so many Degrees to the Southward of the East, and Northward of the West, as in the West Variation; but contrariwise, so many Degrees to the Northwards of the East, and Southwards of the West, as there is East Variation.

A further Use is in many Cases to estimate the Longitude at Sea thereby; for where the Curves run nearly North and South, and are thick together, as about Cape Bona Esperance, it give a very good Indication of the Distance of the Land to Ships come from far; for there the Variation alters a Degree to each two Degrees of Longitude nearly; as may be seen in the Chart. But in this Western Ocean, between Europe and the North America, the Curves lying nearly East and West, cannot be Serviceable for this Purpose.

This Chart, as I said, was made by Observation of the Year 1700, but it must be noted, that there is a perpetual tho' slow Change in the Variation almost every where, which will make it necessary in time to alter the whole System: at present it may suffice to advertise that about C. Bona Esperance, the West Variation encreases at the Rate of about a Degree in 9 years. In our Channel it encreases a Degree in seven Years, but slower the nearer the Equinoctial Line; as on the Guinea Coast a Degree in 11 or 12 Years. On the American side the West Variation alters but little; and the East Variation on the Southern America decreases, the more Southerly the faster; the Line of No Variation moving gradually towards it.

I shall need to say no more about it, but let it command it self, and all knowing Mariners are desired to lend their Assistance and Informations, towards the perfecting of this useful Work. And if by undoubted Observations it be found in any Part defective, the Notes of it will be received with all grateful Acknowledgement, and the Chart corrected accordingly.

E. Halley

This C H A R T is to be sold by William Mount and Thomas Page on Tower-Hill.

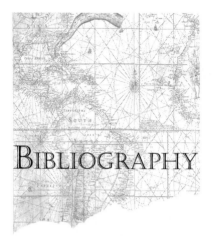

BIBLIOGRAPHY

SELECT SOURCES

The leading account of Halley's three voyages remains *The Three Voyages of Edmond Halley in Paramore 1698-1701* in two volumes. It was published by the Hakluyt Society in London in 1981 and expertly edited by Norman H. W. Thrower.

I also reviewed manuscript journals of Halley's voyages aboard the *Paramore* in the British Library's formidable manuscript room.

Dalrymple's account published in 1773 was the first to include Halley's first and second voyages and provides many insights for modern scholars.

A large portion of Halley's science-based writings have been published in his books and sundry reports in the *Philosophical Transactions* of the Royal Society. Others are found in unpublished papers in the Royal Society's archives in London, which include journal and minute books of the society as well as unpublished notes on presentations to the society.

A top modern source for Halleiana is *Correspondence and Papers of Edmond Halley* by E. F. MacPike, first published in Oxford in 1932. It includes a list of primary sources.

Halley also appears in the diaries and correspondence of other key players of his time. He is mentioned in the diaries of Hooke, Pepys, and Evelyn; in the memoirs of Hearne; and in the letters of Oldenburg, Flamsteed, Newton, Charlett, and Hevelius, among others.

Miscellaneous materials from the National Maritime Museum in Greenwich and its library were also sources for Halley's period. For example, the Navy Board's minutes are archived there.

The National Archives harbors original copies of letters between Halley and Secretary of the Navy Burchett, lords letters to the Navy Board, and treasury papers, among others.

The most comprehensive source for Newton is the *Correspondence of Isaac Newton* in seven volumes, published in Cambridge in 1975.

Unfortunately, one of the best collections of Halley's manuscripts, compiled by former Royal Society Secretary John Machin, was lost. The collection mysteriously disappeared in the 19th century from the library of the Royal Astronomical Society, which is based in London.

The collection of essays assembled to commemorate the tercentenary of Newton's *Principia* and the 1985-1986 return of Comet Halley proved invaluable to properly context Halley's achievements. The essays were published in *Standing on the Shoulders of Giants: A Longer View of Newton and Halley*, which was edited by Norman J. W. Thrower. Its impressive list of contributors includes W. R. Albury, I. Bernard Cohen, Sir Alan H. Cook, Suzanne V. Debarbat, B. J. T. Dobbs, Eric G. Forbes, James E. Force, Gerald Funk, Sara Schechner Genuth, Derek Howse, David W. Hughes, David Kbrin, Simon Schaffer, F. Richard Stephenson, Norman J. W. Thrower, Albert Van Helden, Craig B. Waff, David W. Waters, and Richard S. Westfall.

Very little of Halley's personal correspondence survives.

Unpublished Primary Sources

In the United Kingdom, Cambridge University's King's College Library and the university library house many key documents used to construct events in the narrative. The bulk of John Flamsteed and Halley's papers from their time as royal astronomers at the Royal Greenwich Observatory now reside at Cambridge University's library. So do many pertinent Newton manuscripts.

Outside London the National Archives in Kew houses a plethora of documents going back hundreds of years. The most critical were the Admiralty files, including those from the High Court of Admiralty.

In addition to records of Harrison's court-martial, available papers include the deposition books from 1650 to 1710 and miscellaneous records, including ships' articles, account books, logbooks, apprenticeship indentures, and more. It also holds key documents covering various events in Halley's life, such as the filing of a will and other legal actions.

Although Halley's papers from his days as royal astronomer are now housed at Cambridge, the National Maritime Museum in Greenwich has substantial manuscript holdings pertaining to Halley and his age.

In London the British Library warehouses an impressive collection of Halley's manuscripts and related papers.

The Royal Society's *Collectanea Newtoniana,* Halley manuscripts, and Royal Society journal books, letter books, and council minutes were invaluable to the telling of this story. For example, Halley's original papers on most categories of his work, such as those that would someday lead to the invention of scuba diving, including titles on the diving bell, conveying air into the diving bell, working in the diving bell, and protecting the eardrum while underwater, are all available there for perusing.

The British Museum's exhibit on the Enlightenment, which opened in 2003, also inspired elements of several chapters.

The Paris Observatory's collection of Halley letters and observations helped vivify his visits with Cassini.

Other Printed Sources

Biographia Britannica, 1757, vol. 4, pp. 2494-2520 (London). Either John Machin, a secretary of the Royal Society, or historian Thomas Birch is believed to have authored this. Martin Folkes is thought to have contributed information to the biography.

Eloge by Mairan, printed by MacPike in 1932.

Memoir (anonymous), printed by MacPike in 1931. Some Halley scholars believe that Martin Folkes, then Royal Society president, authored the piece.

Dictionary of National Biography, article on Edmond Halley, by Agnes Clerke. 1949-1950. London: Oxford University Press.

SELECT BIBLIOGRAPHY

"Account of the Comet's Orbit." *The Boston Gazette and Country Journal*, October 1757 to October 1758.

Albury, R. W. 1978. "Halley's Ode on the *Principia* of Newton and the Epicurean Revival in England." *Journal of the History of Ideas*, 29 (1): 24-48.

Allen, J. 1700. *A Full and True Account, of the Behavior, Confessions, and Last Dying-Speeches of the Condemn'd Criminals, That Were Executed at Tyburn, on Friday the 24th May*. London.

Allen, P. 1949. "Scientific Studies in the English Universities of the Seventeenth Century." *Journal of the History of Ideas*, 10 (2): 219-253.

Allibone, T. E. 1973-1974. "Edmond Halley and the Clubs of the Royal Society." *Notes and Records of the Royal Society London*, 28: 195-205.

Alsop, J. D. 1990. "Sea Surgeons, Health and England's Maritime Expansion." *MMI*, xxvi.

Andrewes, W. J. H., ed. 1993. *The Quest for Longitude*. Collection of Historical Scientific Instruments, Harvard University, Cambridge, Mass.

Armitage, A. 1966. *Edmond Halley*. London: Nelson.

Armitage, A., and C. A. Ronan. 1956. "Edmond Halley, 1656-1742." *Memoirs of the British Astronomical Association*, 37 (3).

Armitage, A. *Halley's Astronomical Heritage*.

Atkinson, D. 1999. *Scientific Discourse in Sociohistorical Context: The Philosophical Transactions of the Royal Society of London, 1675-1975*. Mahwah, NJ: Erlbaum Associates.

Aubrey, J. 1958. *Brief Lives*, 3rd ed., O. L. Dick, ed. vol. 1, pp. 282-283, Oxford: Clarendon Press.

Baily, F. 1835. *An Account of the Rev. John Flamsteed*. London: Lords Commissioners of the Admiralty. Reprinted 1966 by Krips Reprint Co., Holland.

Barlow, E. 1934. *Journal of His Life at Sea*, B. Lubbock, ed. Hurst and Blackett. London: F. Muller.

Bassett, F. S. 1885. *Legends and Superstitions of the Sea and Sailors*. New York: Belford Clark.

Bauer, L. A. 1896. "Halley's Earliest Equal Variation Chart." *Terrestrial Magnetism*, (I): 28-31.

Beaglehole, J. C. 1974. *The Life of Captain James Cook*. London: A. and C. Black.

Bennett, J. A. 1975. "Christopher Wren: Astronomy, Architecture, and the Mathematical Sciences." *Journal for the History of Astronomy*, 5 (part 3): 149-184.

Bennett, J. A. 1982. *The Mathematical Science of Christopher Wren*. Cambridge, U.K.: Cambridge University Press.

Berkeley, G. 1734. *The Analyst: An Attack on the Infidel Mathematician Edmond Halley*. London.

Berlinski, D. 2000. *Newton's Gift: How Sir Isaac Newton Unlocked the System of the World*. New York: Free Press.

Birch, T. 1756. *History of the Royal Society of London for Improving Natural Knowledge*, London.

Blane, G. 1785. *Observations on the Diseases of Seaman*.

Boulton, J. 1987. *Neighborhood and Society; a London Suburb in the Seventeenth Century*. London: Cambridge University Press.

Bowen, M. 1929. *Third Mary Stuart, Mary of York, Orange & England; Being a Character Study with Memoirs and Letters of Queen Mary II of England, 1662-1694*. London: John Lane.

Boxer, C.R. 1962. *Golden Age of Brazil, 1695-1750; Growing Pains of a Colonial Society*. Berkeley: University of California Press.

Bradley, J. 1757. "Observations Upon the Comet That Appeared in the Months of September and October 1757, Made at the Royal Observatory." *Philosophical Transactions*, 50: 408-415.

Bram, G., and F. Hogenburg. 1708. *A New View of London*, E. Hatton, ed. London.

Bridge, C. A. G., ed. 1724. *History of the Russian Fleet During the Reign of Peter the Great, by a Contemporary Englishman*.

Bromley, J. S. 1974. "The Manning of the Royal Navy: Selected Public Pamphlets, 1693-1873," *Notes and Records of the Royal Society London*, cxix.

Bromley, J. S. 1987. *Corsairs and Navies, 1660-1760*. London: Hambledon Press.

Brooke, T. H. 1808. *History of the Island of St. Helena, from Its Discovery by the Portuguese to the Year 1806*. London: Black, Perry and Kingsbury.

Brown, P. L. B. 1985. *Halley and His Comet*. Poole, New York: Blanford Press.

Brown, P. L. B. 1986. *Halley's Comet & the Principia*. Aldeburgh, Suffolk, England: Aries Press.

Brown, R. 1981. "The Rise and Fall of the Fleet Marriages" in *Marriage and Society, Studies in the Social History of Marriage*, R. B. Outhwaite, ed., p. 123.

Bullard, E. C. 1956. "Edmond Halley (1656-1741)." *Endeavour*, 15: 189-199.

Bullard, E. C. 1956. "Edmond Halley: The First Geophysicist." *Nature*, 178: 891-892.

Burlingame, A. E. 1969. *The Battle of the Books in Its Historical Setting*. New York: Biblo and Tannen.

Burney, J. 1816. *Chronological History of the Voyages and Discoveries in the South Seas*, vol. 4. London.

Campbell, W. H. 2003. *Introduction to Geomagnetic Fields*. Cambridge, U.K.: Cambridge University Press.

Carrington, A. H. 1939. *Life of Captain Cook*. London: Sidgwick & Jackson.

Chacksfield, K. M. 1988. *1688 Glorious Revolution*. Wincanton, Somerset, England: Wincanton Press.

Chapman, A. 1982. *The Preface to John Flamsteed's "Historia Colestis Britannica."* Greenwich: National Maritime Museum.

Chapman, A. 1994. "Edmond Halley's Use of Historical Evidence in the Advancement of Science." *Notes and Records of the Royal Society London*, 48: 17-191.

Chapman, H. W. 1953. *Mary II, Queen of England.* London: Jonathan.

Chapman, S. 1941. "Edmond Halley as Physical Geographer, and the Story of His Charts." *Occasional Notes of the Royal Astronomical Society*, June 9, 122.

Chapman, S. 1943. "Edmond Halley and Geomagnetism." *Nature*, 152: 231-237.

Clark, D. H., and P. H. Clark. 2001. *Newton's Tyranny: The Suppressed Scientific Discoveries of Stephen Gray and John Flamsteed.* New York: W. H. Freeman.

Clark, G. 1964. *The Later Stuarts (1660-1714).* Oxford: Clarendon Press.

Cohen, I. B., ed. 1958. *Isaac Newton's Papers and Letters on Natural Philosophy.* Cambridge, Mass.: Harvard University Press, pp. 403-404.

Cohen, I. B. 1980. "The Eighteenth-Century Origins of the Concept of Scientific Revolutions." *Journal of the History of Ideas.*

Cohen, I. B. 1987. "Newton's Third Law and Universal Gravity." *Journal of the History of Ideas*, 48 (4): 571-593.

Colledge, J. J. 1969. *Ships of the Royal Navy*, vol. 1. London: Greenhill.

Cook, A. H. 1984. "The Election of Edmond Halley to the Savilian Professorship of Astronomy." *Journal for the History of Astronomy*, 15: 34-36.

Cook, A. H. 1991. "Edmond Halley and Newton's Principia." *Notes and Records of the Royal Society of London*, 45 (2): 129-138.

Cook, A. H. 1993. "Halley the Londoner." *Notes and Records of the Royal Society of London*, 47 (2): 163-177.

Cook, A. H. 1998. *Edmond Halley: Charting the Heavens and the Seas.* Oxford: Clarendon Press.

Cordingly, D. 1995. *Life Among the Pirates: The Romance and the Reality.* New York: Bloomsbury.

Cordingly, D. 1996. *Pirates: Terror on the High Seas, from the Caribbean to the South China Sea.* Atlanta: Turner.

Cotter, C. H. 1972-1973. "A Brief History of the Method of Fixing by Horizontal Angles." *Journal of the Institute Navigation,* 25: 528-534, 26: 491-496.

Cotter, C. H. 1978. "The Mariner's Sextant and the Royal Society." *Notes and Records of the Royal Society of London,* 33: 23-36.

Cotter, C. H. 1981-1982. "Captain Edmond Halley R.N., F.R.S." *Notes and Records of the Royal Society of London,* 36 (1): 61-77.

Cowell, P. H., and A. Crommelin. 1910. *Essay on the Return of Halley's Comet.* Leipzig: W. Engelman.

Cruickshanks, E. 2000. *Glorious Revolution.* London: Macmillan Press.

Culpeper, N. 1653. *The English Physitian [sic] Enlarged.* London: Peter Cole.

Curtin, P. D. 1984. *Cross-Cultural Trade in World History.* New York: Cambridge University Press.

Dalrymple, A. 1773. *Two Voyages Made in 1698, 1699, and 1700, by Dr. Edmund Halley.* London: John Knox.

Davies, J. D. 1992. *Gentlemen and Tarpaulins: The Officers and Men of the Restoration Navy.* Oxford: Clarendon Press.

Deacon, M. 1965. "Founders of Marine Science in Britain: The Work of the Early Fellows of the Royal Society." *Notes and Records of the Royal Society of London,* 20: 28-50.

Deacon, M. 1997. *Scientists at Sea, 1650-1900: A Study of Marine Science.* New York: Academic Press.

Defoe, D. 1727. *Conjugal Lewdness, or Matrimonial Whoredom.* London.

Dick, O. L. 1960. *Aubrey's Brief Lives.* Boston: D. R. Godine

Dodson, J., and W. Mountaine. 1784. *An Account of the Methods Used to Describe Lines on Dr. Halley's Chart of the Terraqueous Globe; Showing the Variation of the Magnetic Needle About the Year 1756, in all the Known Seas; Their Application and Use in Correcting the Longitude at Sea; With Some Occasional Observations Relating Thereto.* London: Mount and Page.

Doyle, W. 1993. *The Old European Order, 1660-1800.* Oxford: Oxford University Press.

Earle, P. 1970. *Corsairs of Malta and Barbary.* Annapolis, Md.: U.S. Naval Institute.

Earle, P. 1989. *The Making of the English Middle Classes: Business, Society and Family Life in London, 1660-1730.* London: Methuen, p. 178.

Earle, P. 1994. *A City Full of People: Men and Women of London, 1650-1750.* London: Methuen.

Earle, P. 1998. *Sailors: English Merchant Seaman, 1650-1775.* London: Methuen.

Ehrman, J. P. W. 1953. *The Navy in the War of William III, 1689-1697. Its State and Direction.* Cambridge, U.K.: Cambridge University Press.

Espinasse, M. M. 1956. *Robert Hooke.* London: William Henemann.

Evelyn, J. 1706. *Sylva.* London.

Evelyn, J. 1950. *Diary and Correspondence of John Evelyn FRS,* W. Bray, ed. London: Henry Colburn.

Fanning, A. E. 1986. *Steady as She Goes: A History of the Compass.* London: Department of Admiralty, National Maritime Museum.

Flamsteed, J. 1995. *Correspondence of John Flamsteed,* E. G. Forbes, L. Murdin, and F. Willmoth, eds. Philadelphia: Institute of Physics.

Flannery, T. 2000. *Terra Australis: Mathew Flinders' Great Adventures in the Circumnavigation of Australia.* Melbourne, Victoria: Text Publications.

Forbes, E. G. 1975. *Greenwich Observatory, vol. 1: Origins and Early History (1675-1835).* London: National Maritime Museum, p. 19.

Fry, H. T. 1970. *Alexander Dalrymple and the Expansion of British Trade.* London: The Royal Commonwealth Society Imperial Studies. No. 29.

Gekkubrabdm, G. 1635. *A Discourse Mathematical of the Variation of the Magneticall Needle Together with Its Admirable Diminution Lately Discovered.* London.

1757. *Gentleman's and London Magazine: And Monthly Chronologer,* 24 (April): 214.

Glass, D. V. 1966. *London Inhabitants Within the Walls 1695.* London: London Record Society.

Glass, D. V. 1968. "Notes on the Demography of London at the End of the Seventeenth Century." *Daedalus,* Spring: xcvii.

Goldsmith, M. M. 1976. "Public Virtue and Private Vices: Bernard Mandeville and English Political Ideologies in the Early Eighteenth Century." *Eighteenth-Century Studies,* 9 (4): 477-510.

Graetzer, J. 1883. *Edmund Halley and C. Neumann (Ein Beitrg zur Gescichte der Bevolkerung-Statistick Beilagen).* Breslau.

Gray, I. 1956. "Peter the Great in England." *History Today,* 225-234.

Green, D. 1970. *Queen Anne.* London: Collins.

Grew, E., and M. Grew. 1910. *Court of William III.* London: Mills and Boon.

Griffiths, A. 1884. *The Chronicles of Newgate.* London: Chapman.

Hall, A. R. 1954. *The Scientific Revolution 1500-1800.* London: Longmans, Green, and Co.

Hampson, N. 1968. *The Enlightenment.* New York: Penguin.

Harris, T., ed. 1990. *Politics of Religion in Restoration England.* Oxford: Basil Blackwell.

Harrison, E. 1696. *Idea Longitudinis.* London.

Hazard, P. 1953. *European Mind, 1680-1715.* London: Hollis & Carter.

Hewson, J. B. 1951. *A History of the Practice of Navigation.* Glasgow: Brown, Son & Ferguson.

Heywood, G. 1994. "Edmond Halley—Actuary." *Quarterly Journal of the Royal Astronomical Society,* 35: 151-154.

Hibbert, C. 1963. *The Roots of Evil: A Social History of Crime and Punishment.* Westport, Conn.: Greenwood Press.

Home, R. W. 1977. "Newtonians and the Theory of the Magnet." *Journal of the History of Science,* 15.

Hooke, R. 1678. *Letters and Collections, Cometa, Microscopium.* London: J. Martyn, pp. 75-77.

Hooke, R. 1707. "A discourse of earthquakes." *Posthumous Works,* R. Waller, ed. London, pp. 270-450.

Hooke, R. 1935. *The Diary of Robert Hooke,* W. H. Robinson and W. A. Adams, eds. London: Taylor & Francis.

Houghton, J. 1727-1728. *Improvement of Husbandry and Trade,* R. Bradley, ed. London.

Howse, D. 1975. *Francis Place and the Early History of the Greenwich Observatory.* New York: Science History Publications.

Howse, D., ed. 1990. *Background to Discovery: Pacific Exploration from Dampier to Cook.* Berkeley: University of California Press.

Howse, D., and M. Sanderson. 1973. *The Sea Chart: An Historical Survey Based on the Collections in the National Maritime Museum.* London: Newton Abbot.

Hunter, M. 1982. *Royal Society and Its Fellows, 1660-1700: The Morphology of an Early Scientific Institution.* Oxford: British Society for the History of Science Monographs.

Huxley, G. L. 1959. "The Mathematical Work of Edmond Halley." *Scripta Mathematica*, 24: 265-273.

Inwood, S. 1998. *A History of London*. London: Macmillan.

Jardine, L. 2003. *The Curious Life of Robert Hooke: The Man Who Measured London*. London: Harper Collins.

Johnson, F. R. 1937. *Astronomical Thought in Renaissance England*. Baltimore, Md.: Octagon.

Jones, Richard Foster. 1961. *Ancients and Moderns: A Study of the Rise of the Scientific Movement in Seventeenth-Century England*. New York: Dover.

Jonkers, A. R. T. 2003. *Earth's Magnetism in the Age of Sail*. Baltimore, Md.: Johns Hopkins University Press.

Kemp, P. 1976. *The Oxford Companion to Ships and the Sea*. Oxford: Oxford University Press.

Kepler, J. 1619. *De Cometis*.

Kilpatrick, C. 1998. *William of Orange: A Dedicated Life, 1650-1702*. Dublin: Education Committee of the Grand Orange Lodge of Ireland.

King, H. C. 1955. *History of the Telescope*. London: Charles Griffin.

King, R. 1994. *Henry Purcell*. London: Thames and Hudson.

Kollerstrom, N. 1985. "Newton's Lunar Mass Error." *Journal of the British Astronomical Association*, 95: 151-153.

Kollerstrom, N. 1992. "The Hollow World of Edmond Halley." *Journal for the History of Astronomy*, 23: 185-192.

Kubrin, D. 1967. "Newton and the Cyclical Cosmos: Providence and the Mechanical Philosophy." *Journal of the History of Ideas*, 28 (3): 325-346.

Kusher, D. 1989. "Secular Acceleration of the Moon's Mean Motion." *Archive for History of the Exact Sciences*, 291-316.

Lambeck, K. 1980. *The Earth's Variable Rotation*. Cambridge, U.K.: Cambridge University Press.

Landers, J., and A. Mouzas. 1988. "Burial Seasonality and the Causes of Death in London 1670-1819." *Population Studies*, 42: 59-83.

Lenfestey, T. 1994. *The Facts on File Dictionary of Nautical Terms*. New York: Facts on File.

Levine, J. H. 1991. *The Battle of the Books*. Ithaca, N.Y.: Cornell University Press.

Levine, J. M. 1977. *Dr. Woodward's Shield: History, Science, and Satire in Augustan England.* Berkeley: University of California Press.

Lillywhite, B. 1963. *London Coffee Houses.* London: George Allen and Unwin.

Lindsay, J. 1978. *The Monster City, Defoe's London, 1688-1730.* London: Grenada Publishing.

Locke, J. 1821. *Two Treatises of Government.* London.

Lorrain, P. 1701. *The Ordinary of Newgate, His Account of the Behavior, Confessions, and Dying-words of Captain W. Kidd, and Other Pirates, That Were Executed . . . May 23, 1701.* London.

Luttrell, N. 1857. *A Brief Historical Relation of State Affairs from September 1678 to 1714.* Oxford.

MacGregor, A., ed. 1994. *Sir Hans Sloane: Collector, Scientist, Antiquary.* London: British Museum Press.

MacPike, E. F. 1902. *Partial Bibliography of Dr. Edmond Halley (1656-1742) with Notes on Other Subjects.* Contributed to the *London Notes and Queries,* 9th series, vols. 10-12. Oxford: Oxford University Press.

MacPike, E. F. 1932. *Correspondence and Papers of Edmond Halley.* Oxford: Clarendon Press.

MacPike, E. F. 1937. *Hevelius, Flamsteed and Halley.* London: Taylor & Francis.

Magrath, J. R. 1921. *Queen's College.* Oxford: Clarendon Press.

Mandeville, B. de. 1724. *An Enquiry into the Causes of the Frequent Executions at Tyburn; And a Proposal for Some Regulations Concerning Felons In Prison.* London.

Marshall, P. J., and G. Williams, 1982. *The Great Map of Mankind: British Perception of the World in the Age of Enlightenment.* Cambridge, Mass.: Harvard University Press.

Masefield, J., ed. 1906. *Dampier's Voyages.* London: Argonaut Press.

Miller, J. 1983. *Glorious Revolution.* New York: Longman.

1695. *Minutes of the Court of Assistants of the Royal African Company.* April 9, 70/84, f. 44r.

Mitchell, T. C., ed. 1979. *Captain Cook and the South Pacific.* London: British Museum Press.

Mohler, N. M., and M. Nicolson. 1937. "The Scientific Background of Swift's 'Voyage to Laputa.'" *Annals of Science,* (2).

Morley, H., ed. 1889. *The Earlier Life and the Chief Earlier Works of Daniel Defoe.* London: George Routledge and Sons.

Murdin, L. 1985. *Under Newton's Shadow.* Bristol and Boston: Adam Hilger.

Norman, R. 1581. *The New Attractive* in Hellman, Gustav, ed. *Rara Magnetica 1268-1599,* Berlin 1898.

Newton, I. 1704. *Opticks.* London: Sam Smith and Benjamin Walford, Printers to the Royal Society.

Newton, I. 1995. *The Principia.* Translated by Andrew Motte. Amherst, N.Y.: Prometheus Books.

Ogg, D. 1955. *England in the Reign of James II and William III.* Oxford: Clarendon Press.

Oliver, S. P. 1880. "Captain Edmond Halley, R.N." *The Observatory,* 3: 349.

Outram, D. 1995. *The Enlightenment.* Cambridge: Cambridge University Press.

Oxford University Statues. 1845. Translated by G. R. Ward. London.

Pool, B. 1973. "Peter the Great on the Thames." *The Mariner's Mirror,* 59: 9-12.

Porter, R. 1994. *London, A Social History.* Cambridge, Mass.: Harvard University Press.

Porter, R. 2000. *Enlightenment: Britain and the Creation of the Modern World.* New York: Palgrave.

Preston, D., and M. Preston. 2004. *A Pirate of Exquisite Mind: Explore, Naturalist and Buccaneer: The Life of William Dampier.* New York: Walker.

1974. *Problems of Medicine at Sea, Maritime Monographs and Reports.* No. 12. London: National Maritime Museum.

Quill, H. 1966. *John Harrison: The Man Who Found Longitude.* London: Baker.

Renier, G. J. 1932. *William of Orange.* Edinburgh: Peter Davies Ltd.

Rice, T. 1999. *Voyages of Discovery: Three Centuries of Natural History Exploration.* London: Scriptum Editions in Association with the Natural History Museum.

Rigaud, S. J. 1844. *A Defence of Halley Against the Charge of Religious Infidelity.* Oxford: Asmolean Society.

Riley, J. C. 1981. "Mortality on Long-Distance Voyages in the Eighteenth Century." *Journal of Economic History,* 41(3): 651-656.

Ritche, R. C. 1986. *Captain Kidd and the War Against the Pirates.* Cambridge, Mass.: Harvard University Press.

Robert, R. 1969. *Chartered Companies and Their Role in the Development of Overseas Trade.* London: Bell.

Ronan, C. A. 1968. "Edmond Halley and Early Geophysics." *Geophysical Journal,* 15: 241-248.

Ronan, C. A. 1969. *Edmond Halley: Genius in Eclipse.* New York: Doubleday.

Ronan, C. A. *Edmond Halley; The Man and His Work.*

Royal Society. 1837. *Collectanea Newtoniana, March 1693 to January 1702,* IV(4): 18-72.

Rudolph, A. H. 1904. "Material for a Bibliography of Dr. Edmond Halley." *Bulletin of Biography,* 4: 54-57.

Salmon, W. 1693. *The Compleat English Physician: Or the Druggists' Shop Opened.* London: printed for Mathew Gulliflower at the Black Speak Eagle in Westminster-Hall.

Sanz, C. 1964. *Australia: Its Discovery and Name; with Facsimile Reproductions of the Quieros Memorial and Other Rare Illustrations.* Madrid: Direccion General de Relccines Culturales (Lond 1939).

Scott, J. F., ed. 1967. *The Correspondence of Isaac Newton.* Cambridge, U.K.: Cambridge University Press.

Seller, J. 1669. *Practical Navigation.* London: British Library Collection.

Seller, J. 1677. *The English Pilot.* London: British Library Collection.

Shapiro, B. J. 1983. *Probability and Certainty in 17th-Century England; A Study of the Relationships Between Natural Science, Religion, History, Law and Literature.* Princeton, N.J.: Princeton University Press.

Shroeder, W., ed. 2000. *Geomagnetism Research: Past and Present.* Darmstadt, Germany: International Association of Geomagnetism and Aeronomy.

Sloane, K., ed. 2003. *Enlightenment: Discovering the World in the 18th Century.* London: The British Museum Press.

Smith, H. 1994. *English Channel: A Celebration of the Channel's Role in England's History.* Upton-Upon-Severn, Worcestershire, England: Images Publications.

Sobel, D. 1995. *Longitude: The True Story of a Lone Genius Who Solved the Greatest Scientific Problem of His Time.* New York: Penguin.

Somner, W. 1693. *A Treatise of the Roman Ports and Forts in Kent, Early English Books, 1641-1700.* London: James Brome, pp. 194-204.

Stephens, E. 1689. *Reflections Upon the Occurrences of the Past Year.* London.

Stone, L. 1900. *The Family, Sex and Marriage in England 1500-1800.* New York: Harper and Row.

Stow, J. 1720. *A Survey of the Cities of London and Westminster . . . Brought Down from the Year 1633 . . . to the Present Time by John Strype.* London.

Strong, E. W. 1952. "Newton and God." *Journal of the History of Ideas,* 13 (2): 147-167.

Swift, J. 1990. "Dean of St. Patrick's." In *Gulliver's Travels and Selected Writings in Prose and Verse,* J. Hayward, ed. Oxford: Oxford University Press.

Tafel, R. L. 1875-1877. *Documents Concerning the Life and Character of Emanuel Swedenborg.* London: Swedenborg Society.

Tanner, J. R., ed. 1926. *Samuel Pepys's Naval Minutes.* Vol. 60. London: Naval Records Society.

Taylor, E. G. R. 1962. *Geometrical Seaman: A Book of Early Nautical Instruments.* London: Institute of Navigation.

Taylor, E. G. R. 1971. *The Haven-Finding Art: A History of Navigation from Odysseus to Captain Cook.* London: Institute of Navigation.

Thrower, N. J. W. 1969. "Edmond Halley and Thematic GeoCartography," *The Terraqueous Globe.* Los Angeles: William Andrews Clark Memorial Library.

Thrower, N. J. W., ed. 1981. *The Three Voyages of Edmond Halley in the "Paramore," 1698-1701.* London: Hakluyt Society Publications.

Thrower, N. J. W., ed. 1990. *Standing on the Shoulders of Giants: A Longer View of Newton and Halley.* Berkeley: University of California Press.

Thrower, N. J. W. 1996. *Maps and Civilization: Cartography in Culture and Society,* 2nd ed. Chicago: University of Chicago Press.

Tinkler, J. F. 1988. "The Splitting of Humanism: Bentley, Swift, and the English Battle of the Books." *Journal of the History of Ideas,* 49 (3): 453-472.

Turnbull, W. H., and J. F. Scott. 1960. *Correspondence of Isaac Newton.* Cambridge.

Tuveson, E. 1950. "Swift and the World-Makers." *Journal of the History of Ideas*, 11(1): 54-74.

Van der Zee, H., and W. 1973. *William and Mary*. London: Macmillan.

Verschuur, G. L. 1993. *Hidden Attraction: The History and Mystery of Magnetism*. Oxford.

Waff, C. B. 1986. "Comet Halley's First Expected Return: English Public Apprehensions." *Journal for the History of Astronomy*, 17: 1-37.

Walker, R. 1794. *A Treatise on Magnetism, with a Description and Explanation of a Meridional and Azimuth Compass*. New York: Academic, pp. 165-192.

Waller, M. 2000. *1700: Scenes from London Life*. New York: Four Walls Eight Windows.

Waters, D. W. 1958. *The Art of Navigation in England in Elizabethan and Early Stuart Times*. New Haven, Conn.: Yale University Press.

Weinreb, B., and C. Hibbert, eds. 1986. *London Encyclopedia*. New York: St. Martin's Press.

West, R. 1997. *The Life and Surprising Adventures of Daniel Defoe*. London: Carrol and Graf.

Westfall, R. S. 1980. *Never at Rest: A Biography of Isaac Newton*. New York: Cambridge University Press.

Whiston, W. 1714. *The Cause of the Deluge Demonstrated*. London.

Whiston, W. 1753. *Memoirs of the Life and Writings of Mr. William Whiston*. London.

Whiston, W., and H. Ditton. 1714. *A New Method for Discovering the Longitude*. London.

Williams, J. E. D. 1992. *From Sails to Satellites: The Origin and Development of Navigational Science*. Oxford: Oxford University Press.

Willmoth, F. 1993. *Sir Jonas Moore: Practical Mathematics and Restoration Science*. Woodbridge, Suffolk: Boydwell Press.

Woolley, R. 1969. "Captain Cook and the Transit of Venus of 1769." *Notes and Records of the Royal Society of London*, 24: 19-32.

Wrigley, E. A. 1967. "A Simple Model of London's Importance 1650-1750." *Past and Present*, 37.

Yeomans, D. K. 1982. *The Comet Halley Handbook*. Washington, D.C.: Jet Propulsion Lab Publications.

A sincere thanks to Dimas Ayala Riquelme at the International Monetary Fund for assisting with the conversion of pounds of Halley's time to U.S. dollars of today.

I also thank John Hwang, Christopher Johnson, Eric Maxeiner, and Scott Kamp for serving as test readers of my early chapters; Kim Monk, Valerie Jablow, and Bill Horne for reading the almost finished chapters.

I am also appreciative of all the talented editors who taught me to write on the job over the years and who gave me fantastic writing opportunities: Eric Bates, J. Kelly Beatty, Frank Best, Susan M. Booker, James Burkhart, David Carr, John Cloud, Sherri Dalphonse, Sean Daly, Ken DeCell, Glenn Dixon, Alexis Doster, David Eicher, Timothy Foote, Paul Glastris, Bill Goggins, Alex Heard, Valerie Jablow, Clara Jeffery, Jack Limpert, James Lockhart, Colin Macilwain, Sally Scott Maran, Stephanie Mencimer, Don Moser, Rachel Nowak, Daniel Pendick, David Plotz, Miles Porter, Ghassan Rassam, John Rennie, Jack Shafer, Gary Stix, Clint Talbott, Kimberly G. Thigpen, Nancy Tomich, Debra Melani, Gabrielle Walker, Erik Wemple, Catherine White, John P. Wiley, Jr., Carey Winfrey, Philip Yam, and Glenn Zorpette.

I am especially thankful for the support of my wonderful friends and family throughout this project.

Washington, D.C.
January 2005

J.W.

Works by Halley

"An Account of the Appearance of an Extraordinary Iris Seen at Chester in August Last," *Philosophical Transactions,* 20: 193-196 (1698).

"An Account of the Cause of the Change of the Variation of the Magnetical Needle; with an Hypothesis of the Structure of the Unternal Parts of the Earth," *Philosophical Transactions,* 17: 563-578 (1692).

"An Account of the Late Surprizing Appearance of the Lights Seen in the Air, on the Sixth of March Last, with an Attempt to Explain the Principal Phaenomena Thereof," *Philosphical Transactions,* 406-428: 427 (1716).

"Astronomiae Cometicae Synopsis," *Philosophical Transactions,* 24: 1882-1899 (1704-1705).

"An Estimate of the Degrees of Mortality of Mankind; Drawn from Curious Tales of the Births and Funerals at the City of Breslaw; with an Attempt to Ascertain the Prices of Annuities Upon Lives," *Philosophical Transactions,* 17(196): 596-610, 654-656 (1693).

"Farther Thoughts on the Same Subject," *Philosophical Transactions.* No. 383 (1724).

"An Historical Account of the Trade Winds, and Monsoons, Observable in the Seas Between and Near the Tropicks, with an Attempt to Assign the Physical Cause of the Said Winds," *Philosophical Transactions,* (16): 153-168 (1686).

"An Instance of . . . Finding the Focie of Optick Glasses Universally," *Philosophical Transactions,* 17: 960-969 (1691-1693).

Letter, Halley to Hooke, St. Helena, November 22, 1677, with an observation on the Transit of Mercury.

"A Method of Enabling a Ship to Carry Its Guns in Bad Weather," in E. F. MacPike, ed., pp. 164-165 (1932).

"Monsieur Cassini His New and Exact Tables for the Eclipses of the First Satellite of Jupiter, Reduced to the Julian Stile, and Meridian of London," *Philosophical Transactions,* 18 (214): 237-256 (1694).

"A New and Correct Chart Showing the Variations of the Compass in the Western and Southern Oceans as Observed in Ye Year 1700 by his Maties Command by Ed. Halley." (1701). London: Mount and Page.

"Observations Made on the Eclipse of the Moon, on March 15, 1735/6, (1737-8)." *Philosophical Transactions,* 40: 14 (1737-1738).

"A Proposal for a Method for Finding the Longitude at Sea Within a Degree or 20 Leagues," *Philosophical Transactions*, 37: 185 (1731).

"A Short Account of the Cause of the Saltiness of the Oceans, and of the Several Lakes That Emit No Rivers, with a Proposal by Help Thereof, to Discover the Age of the World," *Philosophical Transactions*, 29 (244): 296-300 (1715).

"Some Account of the Ancient State of the City of Palmyra with Short Remarks Upon the Inscriptions Found There," *Philosophical Transactions*, 218: 160-175 (1695).

"Some Consideration About the Cause of the Universal Deluge," *Philosophical Transactions*. No. 383 (1724).

"Some Further Consideration on the Breslaw Bills of Mortality. By the Same Hand," *Philosophical Transactions*, 17 (197): 654-656 (1693).

"Some Remarks on the Variations of the Magnetical Compass (1714-16)," *Philosophical Transactions*, 14: 165-168 (1684).

"A Theory of Tides at the Bar of Tunkin," *Philosophical Transactions*, 14: 685-688 (1684).

"A Theory of the Variation of the Magnetical Compass," *Philosophical Transactions*, 13 (148): 208-221 (1683).

"The True Theory of the Tides, Extracted from That Admired Treatise of Mr. Isaac Newton, intitled *Philosophiae Naturalis Principia Mathematica*, Being a Discourse Presented with that Book to the Late King James," *Philosophical Transactions*, 19 (226): 445-457 (1697).

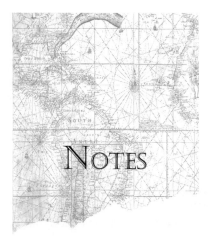

NOTES

CHAPTER 1
CAST OFF

All dates in this book have been adjusted. England was still adhering to the Julian calendar throughout Halley's life. The monarchy did not adopt the Gregorian calendar until 1752. It was viewed as popish superstition. As a result, dates in England prior to 1700 were 10 days later than those of the continent. England also observed 1700 as a leap year, so after 1700 the calendars were out of sync by 11 days. To further confuse things, England began the year on March 25.

A full account of Halley's voyages did not appear in print until 1775 with the publication of Dalrymple's *A Collection of Voyages Chiefly in the Southern Atlantic Ocean.*

The correct pronunciation of Halley's name is a subject of perennial debate. Whether Edmond himself pronounced it "Hawley" or "Hali" or "Haili" is impossible to know for certain. His name is also spelled different ways in records and correspondence, including Hailey, Haley, Haly, Hawley, and Hally. It was listed on his marriage certificate and on his final will and testament, drawn up in June 1736, as Edmond Halley. When he signed his full signature, which was rarely, he used this spelling as well. Because he often abbreviated his first name as "Edm.," it is often in essence misspelled "Edmund."

Halley was never knighted, so his proper title was "Dr.," not "Sir." He received an honorary degree of doctor of laws from Oxford in 1710.

Scholars know little of Benjamin Middleton, the son of Colonel Middleton, who helped Halley organize the mission at the outset and may have played an important role politically in gaining the queen's support. He clearly was an experienced seaman but apparently was not involved with Halley's mission after the initial conception of the plan. Originally, he offered to finance the crew and supplies for the mission, but the monarchy agreed to pay for not only the vessel and its arms but to "victual and man" the *Paramore*. The Admiralty also built a larger ship than the 60-ton vessel Middleton first requested. Both Middleton and his servant Sir John Hoskins were to accompany Halley on the voyage, but neither is ever mentioned again after the earliest stages of preparation.

Although the Royal Society was officially established in 1660, it had met on a pro forma or ad hoc basis since roughly 1645 under the name of the *Philosophical or Invisible College*. Its appellation as the *Royal Society of London for Improving Natural Knowledge* did not appear in print until 1661 and until 1663 in the second Royal Society charter. The Royal Society has existed ever since. It was essentially the first public institution for the pursuit of scientific research.

Samuel Pepys remarked on Halley's navigation talents in his *Naval Minutes*, edited by J. B. Tanner and published in 1926.

Sydney Chapman claims that Halley's mission was "the first voyage undertaken for a purely scientific object," in his 1941 paper published in *Occasional Notes No. 9, Royal Astronomical Society*. His claim is based on the fact that he was the first renowned scientist to plan and execute a voyage to test his hypothesis that was fully funded by a government and not for commercial advantage. Likewise, in 1849 Alexander von Humboldt considered the voyages thusly in his *Cosmos: A Sketch of a Physical Description of the Universe*: "Never before, I believe, had any government fitted a naval expedition for an object whose attainment promised such advantages to practical navigation, while at the same time it deserved to be regarded as peculiarly scientific and physico-mathematical." Similarly, in 1849, Captain S. P. Oliver stated, according to *The Observatory*: "We do not often think of him as a sailor; and yet, previous to Cook, Captain E. Halley was our first scientific voyager." Moreover, for many years after his death Halley was touted by his contemporaries as "the second [after Newton] most illustrious of the Anglo-Saxon Philosophers," according to an 1880 issue of *Nature*. Although there were earlier English expeditions, they were not true scientific voyages.

Societies for other disciplines besides science would emerge in the coming century. In the early 1700s, societies for Antiquaries, Dilettanti, the Encouragement of Arts, Manufactures and Commerce, Asian studies, and more would form. Such societies and clubs provided a public forum and captive audience that when paired with the British concept of public spirit created enthusiasm for new ideas like never before.

For more on classes of sea captains, see J. D. Davies's *Gentlemen and Tarpaulins: The Officers and Men of the Restoration Navy*. As Halley biographer MacPike put it: "The idea of commissioning a landsman to the command of a Kings' ship might appear to professional seamen as sheer madness.... However, in Halley's case the rash outrageous act was justified in the event, but only because Edmond Halley was one man in a thousand, possess[ed] of the most varied gifts and the most extraordinary versatility."

Halley's grandfather, Humphrey Halley, was a vintner and a haberdasher. Like assorted dynasties, the manufacture of alcohol contributed to establishing his family's fortune.

The description of Mary's funeral is given in Henri and Barbara van der Zee's *William and Mary*.

Accounts of the Glorious Revolution can be found in books by that name written by Eveline Cruickshanks, John Miller, and K. Merle Chacksfield.

The punctuation, grammar, spellings, etc., in many direct quotations throughout the narrative have been modernized for clarity. Some original spellings were kept to capture the flavor of the time.

CHAPTER 2
THE ALMOST-LOVABLE *PARAMORE*

There was only one ship ever named the *Paramore* in the Royal Navy. Halley consistently spelled it that way. However, sundry other spellings are found in other correspondence and records regarding the ship, including *Paramour, Parrimore,* and *Parramore.* She was clearly the first ship built by the Royal Navy solely for a science mission. See T. D. Manning and C. F. Walker in *British Warship Names,* published in London in 1959.

For an idea of the lawlessness of life at sea, sea William Byam's account published in London in 1665. It is titled, "Exact Relation of the Most Execrable Attempts of John Allin, Committed on the Person of His Excellency Francis Lord Willoughby of Parham, Captain General of the Continent of Guiana, and of All the Caribby-Islands, and Our Lord Proprietor."

Some simply define navigation as "finding the way."

Teredo worms, which reach up to six inches in length, are a variety of mollusk that reproduce faster than rats on a ship. One worm can lay a million eggs a year.

For a full account of how the phenomenon of magnetic variation put Christopher Columbus off course, try Samuel Elliot Morrison's *Admiral of the Ocean Sea,* published by Little, Brown, in 1942.

For an outstanding account of the history of astronomical methods to determine longitude at sea, see Charles Cotter's *A History of Nautical Astronomy,* pp. 180-267.

CHAPTER 3
BATTLE OF THE BOOKS

Tradition holds that debate over the origins of an embossed iron buckler may have ignited the literary Battle of the Books. Its owner, Cambridge physician and collector Dr. John Woodward believed it to be a Roman shield dating to the Gaul invasion of Rome. It proved to be made in France around 1540. For more information, see J. M. Levines's *Dr. Woodward's Shield.*

There are many books on the Battle of the Books that offer widely ranging views on its meaning and significance. The 1961 study by R. F. Jones of the rise of the scientific movement in 17th-century England entitled *Ancients and Moderns* is especially relevant.

Likewise, a host of titles cover the Enlightenment, including both D.

Outram's and Hampson's eponymous *The Enlightenment* and R. Porter's *Enlightenment: Britain and the Creation of the Modern World.*

For more on the intelligentsia's hangouts, see *London Coffee Houses: A Reference Book of Coffee Houses of the Seventeenth, Eighteenth, and Nineteenth Centuries* and B. Lillywhite's *London Coffee Houses.*

Good sources on the history of the Royal Society include Dwight Atkinson's *Scientific Discourse in Sociohistorical Context: The Philosophical Transactions of the Royal Society of London, 1675-1975,* and Michael Hunter's *Royal Society and Its Fellows, 1660-1700, the Morphology of an Early Scientific Institution.* Its governing body was a committee of 23 men, including an elected president, vice president, and two secretaries. During its early decades a single Secretary had the responsibility for publishing its *Philosophical Transactions.*

The first problem the French Academie des Sciences tackled was determining the precise length of the degree of the meridian. Under the direction of King Louis XIV's powerful minister Jean Baptiste Colbert, the project was undertaken in 1669 by Jean Picard and several of his associates and completed in 1770. Picard used triangulation to solve the problem. While the Royal Society pushed its then-curator Robert Hooke to make the measurement, it was too daunting a task for an individual effort. Colbert is said to have envisioned a society that was more a factory than a marketplace of ideas.

Some scholars suggest that Robert Boyle also may have supported Halley's first venture to St. Helena. He was an active Royal Society fellow and a director of the East India Company from 1662 through 1677.

John Caswell was another candidate for the Savilian Chair of Astronomy in 1691. Gregory, of course, was successful, but on his death in 1708, Caswell, a friend of Flamsteed's, was then appointed. If an Englishman was selected, he had to hold a master's degree. The duties of the Savilian professors were specific but very liberal:

> The professor of Geometry must understand that it is his proper province publicly to expound the thirteen books of Euclid's Elements, the Conies of Apollonius, and all the books of Archimedes.... However, as to undertaking, or not, the explanation of the Spherics of Theodosius and Menelaus, and the doctrine of Triangles, as well plane as spherical, I

leave the option at large to both professors. It will besides be the business of the Geometry professor, at his own times . . . to teach and
expound arithmetic of all kinds, both speculative and practical; land-
surveying or practical geometry; canonics of music and mechanics. And
in explaining the above departments, I leave the professor a free choice
of the books which he chooses to explain, unless the University think
otherwise.

Geometry, in this sense, included a range of mathematics and essentially what is considered physics today. Meanwhile, the astronomy professor was to "explain the whole of the mathematical economy of Ptolemy
(usually called the Almagest), applying in their proper place the discoveries
of Copernicus, Gexber, and other modern writers . . . the whole science of
optics, gnomonies, geography, and the rules of navigation in so far as they
are dependent on mathematics. He must understand, however, that he is
utterly debarred from professing the doctrine of nativities and all judicial
astrology without exception." Two years later in 1621, Savile's son-in-law,
William Sedley, followed suit and endowed a natural philosophy chair.

Oxford's first two astronomy professors, John Bainbridge and John
Greaves, subscribed heavily to Ptolemy's views, so the Copernican system
was not promoted until almost 1649 when Seth Ward assumed the chair.
Oxford's first professors in botany and chemistry would be set up in 1669
and 1683, respectively.

John Wallis explained his independent study of math in this way: "I
had none to direct me, what books to read, or what to seek, or in what
Method to proceed. For Mathematics (at that time, with us) were scarce
looked upon as academical studies, but rather mechanical; as the business
of traders, merchants, seamen, carpenters, surveyors of lands, or the like;
and perhaps some almanac-makers in London. And amongst more than
two hundred students (at that time) in our college, I do not know of any
two (perhaps not any) who had more of Mathematics than I (if so much)
which was then but little. . . . For the study of Mathematics was at that time
more cultivated in London than in the Universities," according to Allen.

In fact, the first professorship promoting the so-called new philosophy at Cambridge was established in 1663 by Henry Lucas in mathematics.
Isaac Barrow was selected as the first Lucasian professor. Not surprisingly,
the universities initially resented the establishment of the Royal Society

and its push to grant degrees. Oxford, in fact, thought the society "obnoxious," according to Anthony Wood. Newton took over in 1669. Once the Royal Society was founded, there was somewhat of an exodus of science minds from Oxford to London. Wren, Hooke, and Boyle joined Ward, Wilkins, Goddard, and Rooke there. As a result, Cambridge became the leading refuge for the "new philosophy" toward the end of the 17th century. However, it was not the most modern institution in England. The Restoration Parliament passed the Act of Uniformity in 1662, which, among other things, mandated orthodoxy to the Church of England for enrollment, excluding so-called nonconformists and sparking the establishment of other institutions of higher learning.

For a good understanding of the system of patronage that existed in Halley's day, the first chapter of *Standing on the Shoulders of Giants*, by Richard S. Westfall and Gerald Funk, gives a wonderful analysis. Titled "Newton, Halley, and the System of Patronage," it explains the nature and distinguishes Newton's relationship to both Halley and Gregory.

To learn more about the history of the Royal Observatory, see *Greenwich Observatory . . . The Story of Britain's Oldest Scientific Institution, the Royal Observatory at Greenwich and Herstmonceux, 1675-1975*. A castle used to stand on the site, which is the highest point of elevation in Greenwich's Royal Park. For more on the observatory's founding, see F. Willmoth's *Sir Jonas Moore* and the introduction to *Flamsteed's Correspondence*. Derek Howse's chapter, "Newton, Halley, and the Royal Observatory," in *Standing on the Shoulders of Giants*, is also very thorough.

For more on William Whiston's work, see *The Cause of the Deluge Demonstrated and Memoirs of the Life and Writings of Mr. William Whiston*.

CHAPTER 4
TROUBLE ON THE PINK

For more on Dampier, see his *New Voyage Round the World*, with an introduction by Sir Albert Gray. Diana and Michael Preston's *A Pirate of Exquisite Mind: Explorer, Naturalist, and Buccaneer: The Life of William Dampier* is a solid, contemporary retelling of his adventures.

For a look at other scientists' impact on the evolution of marine research, Margaret Deacon offers a thorough overview in her 1997 work, *Scientists at Sea, 1650-1900: A Study of Marine Science*.

A deeper look at the origins of the study of natural history can be found in T. Rice's *Voyages of Discovery: Three Centuries of Natural History Exploration.*

For more on the development of the compass and the azimuth compass, in particular, see Jonkers's chapter, "Following the Iron Arrow," in his *Earth's Magnetism in the Age of Sail.* The azimuth observation was the most difficult of the compass readings. Technically, it is the horizontal arc between the local meridian and the vertical plan through the sighted object and the observer.

Needle deflection that is attributable to nearby iron is called magnetic deviation. The magnetic properties of hard and soft iron cause such variation. Hard iron is not readily magnetized but retains its magnetism permanently. Soft iron is easily realigned with passing magnetic fields. All iron materials possess both hard and soft properties but to varying degrees, which determine their overall magnetization.

Portugal's Pedro Nunez was the first to explain that, if the Sun's height and declination were known, its amplitude, that's the angle from due east or west that the Sun rises or sets, could be found by spherical trigonometry.

In 1677, Pepys, in his role as Admiralty secretary implemented the naval lieutenant's examination. The policy served to promote mathematical skill among naval officers and navigators.

Knowing the "place of the Moon," relative to the Sun or stars at a given moment, was required to use the lunar distance method to determine longitude. Throughout Halley's lifetime, it would remain an impossible feat. But in the 1760s it became a viable option on the basis of Newton's theory of the Moon, which was published in David Gregory's *Astronomiae Physicae* in 1702.

Halley's approach was based on the saros or eclipse cycle of 18 years and 11 days or that the relative position of Sun and Moon repeats every saros, that is, every 223 lunations. In this way the place of the Moon could be predicted relative to the Sun by looking at its observed location 18 or 36 years earlier. On his Atlantic expeditions, Halley managed to use his method successfully by observing occultation and appulses of stars by the Moon.

For a detailed and well-researched account on the life of sailors during this time, see Peter Earle's *Sailors.* I am indebted to Earle's research on this topic.

There are a flurry of interesting pirate books out there. David Cordingly has authored and edited several, including *Life Among the Pirates: The Romance and the Reality* and *Pirates Terror on the High Seas, from the Caribbean to the South China Sea.*

To learn more about the Corsairs, see the *Corsairs of Malta and Barbary,* published in 1970 also by Peter Earle.

Halley described the polypus in the May 1, 1689, *Journal Books of the Royal Society,* which was extracted by H. W. Robinson.

For more on the politics of Brazil, see the *Golden Age of Brazil, 1695-1750, Growing Pains of a Colonial Society.*

James Burney described the tarpaulin attitude well in his 1816 *Chronological History of the Voyages and Discoveries in the South Seas.* He writes: "Respect for science, however, did not operate sufficiently strong on the Officers of Dr, or rather Captain, Halley's ship, to prevent their taking offence at being put under the command of a man who had risen without going through the regular course of service with the Royal Navy."

Through the 19th century, studies at Oxford University were governed by the Laudian Statutes of 1636, which is also known as the Caroline Code. Since Archbishop Laud was a member of the clergy, he emphasized theological studies. The rigor of his regulations hindered the study of science as they were chiefly intended to bolster education in terms of the tenets of the Royalists and the High Church.

To earn a master's at Oxford, the average student had to study three additional years, or 12 terms, completing courses in Greek, which covered Homer, Demoshtes, Isocrates, and Euripides; Aristotelian metaphysics; more geometry, astronomy, natural philosophy and Hebrew, using the Bible and the work of Lucius Florus and other ancients, according to Phyllis Allen's interpretation in the *Journal of the History of Ideas.*

CHAPTER 5
FRIENDSHIP ROYAL

The term "round robin" first appeared in print in England in 1546 in reference to the sacrament of communion. Calvin used the term derisively in the context of blasphemous practices, according to Words@Random.

For an analysis of Halley's introduction to Newton's *Principia,* see I.

Bernard Cohen's "Halley's Two Essays on Newton's *Principia*" in *Standing on the Shoulders of Giants.*

To learn about Newton's surprising ideas about alchemy, check out B. J. T. Dobbs in "Newton as Alchemist and Theologian" in *Standing on the Shoulders of Giants.*

Paul Lorrain did a comprehensive job of chronicling events at Newgate in the early 18th century.

After Lieutenant Edward Harrison's court-martial proceedings, he offered the data he had collected, which at times differed from Halley's observations, to John Flamsteed. Harrison was well aware of their rivalry. Flamsteed was interested and had Harrison's journal copied in March 1700, according to Flamsteed's correspondence.

Dampier would establish his place as a seafarer in other ways. On a subsequent privateering expedition in 1704, Dampier stranded an unruly Scottish sailor named Alexander Selkirk at his request on the island of Juan Fernandez off the coast of Chile. The incident inspired Daniel Defoe's *Robinson Crusoe.* Dampier wrote extensive memoirs of his voyages in which he had only praise for Halley and his quest: "I cannot but hope that the ingenuous Author, Captain Halley, who to his profound skill in all theories of these kinds hath added and is adding continually personal experiments, will ever long oblige the world with a fuller discovery of the course of the variations, which hath hitherto been a secret."

CHAPTER 6
OUTWARD BOUND

In his appendix to the 1710 edition of English scholar Thomas Streete's *Astronimia Carolina*, Halley noted that practice observations from the deck of a moving ship were possible in mild conditions with telescopes of five to six feet in length.

For more on the role of the eclipses of Jupiter's moons, see Albert Van Helden's contribution, "Longitude and the Satellites of Jupiter," to the *Quest for Longitude.*

To put navigation in better perspective, see J. E. D. Williams's *From Sails to Satellites: The Origin and Development of Navigational Science* for a worthwhile read.

CHAPTER 7
TERRA INCOGNITA

In Paris, Cassini and his colleague Jean Richer had estimated in 1672 the distance between the Sun and Earth using observations of Mars at particular positions. Based on observations from Paris and French Guiana, he calculated the AU to be 87 million miles. Though smaller than the real AU, its validity wasn't widely acknowledged until the Venus work was completed.

At this time the Union flag bore St. George's Cross of England on St. Andrew's Cross of Scotland, which had been in use since 1603 when the crowns were united. This happened after Queen Elizabeth I died without an heir. James VI of Scotland became James I of England, marking the start of the Stuart dynasty in England.

In 1697 William Dampier sailed to Australia in the pirate ship *Cygnet* and made the first landing by an English explorer in Australia. The spot where he came ashore is to be renamed Dampier's Landing, in his honor. It is likely that on this voyage he collected two plant specimens, an *Acacia* and *Synaphea*. Two years later, in 1699, sailing in HMS *Roebuck*, Dampier landed on Dirk Hartog Island in western Australia and made a collection of specimens of flora as well as drawings of birds, fish, and other animals of this New World.

To commemorate the tri-centenary of the voyage, Dampier's collection of 24 plant specimens was exhibited at the Museum of Western Australia, on loan from England's Oxford Herbarium. The *Roebuck* was shipwrecked on the voyage back to England, but miraculously Dampier's specimens survived. An account of this voyage is found in his book, *A Voyage to New Holland*. The well-known story *Gulliver's Travels*, by Jonathan Swift, was based partly on Dampier's travels.

A good source on Cook's expeditions is T. C. Mitchell's *Captain Cook and the South Pacific*.

CHAPTER 8
COMPASS POINTS

Halley described his diving engine in the March 6, 1988-1989, *Journal Book of the Royal Society*, extracted by H. W. Robinson.

For more on the medical science of the times, see Maureen Wallers's *1700: Scenes from London Life*.

The first meteorological diagram was published in 1684 in the *Philosophical Transactions of the Royal Society in London* (no. 169), based on barometric observations made at Oxford, according to G. Hellman, a renowned German historian of science.

See *History of the Russian Fleet During the Reign of Peter the Great, by a Contemporary Englishman* to follow up on Peter's interests in building his naval forces.

David W. Waters's "Captain Edmond Halley, F.R.S. Royal Navy, and the Practice of Navigation," in *Standing on the Shoulders of Giants*, provides more details on Halley's contributions in this arena.

CHAPTER 9
CHART THE NEEDLE

For help explaining the basics of geomagnetism, Wallace H. Campbell's *Introduction to Geomagnetic Fields* was very useful.

Some scholars credit Robert Norman, an English seaman and compass crafter, for inspiring Gilbert's *De Magnete*. In 1581, Norman published *The New Attractive* in London. In it, using science-based methods, he details his discovery of the magnetic dip.

For a contemporary synthesis of the relationship between magnetism and navigation, see A. R. T. Jonkers's *Earth's Magnetism in the Age of Sail*. Jonkers is more skeptical of Halley's contributions than many other scholars.

Two of Halley's four magnetic poles were proven not to exist in 1817, when a complete chart of magnetic meridians was first published. This is explained in full in an article by Sydney Chapman that appeared in *Nature* in 1943.

According to MacPikes's *Hevelius, Flamsteed, and Halley*, no serious scientist has ever taken seriously Flamsteed's claims that Halley borrowed ideas from Perkins's theory. However, in a December 12, 1700, letter to Flamsteed, Perkin's brother, Thomas Perkins, stated that Halley bought his papers two years earlier, soon after his brother's death.

Norman J. W. Thrower does an excellent job of placing Halley's contributions to cartography into historical perspective in his recent *Maps & Civilization: Cartography in Culture and Society*.

Halley apparently believed that Mercator didn't deserve as much credit as he received for developing the method of map projection that emerged in 1569. For this reason, Halley proposed the name "nautical" for when the charts were used by navigators. Mercator, however, stuck.

Norman J. W. Thrower's chapter, "Longitude in the Context of Cartography," in the *Quest for Longitude* goes into more detail.

Alexander von Humbolt first proposed the words "isogonic," "isoclinic," and "isodynamic" to describe lines of equal variation, dip, and magnetic field strength, respectively, according to Charles Cotter.

Two other manuscripts exist that detail isobaths, or equal depths of water. Although Halley was unaware of their existence, they were published in 1584 and 1697, respectively. But none published lines of equal declination.

For a complete biography on Sloane, try A. MacGregors's *Sir Hans Sloane: Collector, Scientist, Antiquary*.

Researchers have branded magnetic devices that supposedly alleviate pain as "nonsense." According to the "Wellness Letter," published by the University of California at Berkeley School of Public Health, "There is no good scientific evidence—or any logical reason to believe—that magnets can relieve pain."

For more commentary on his Atlantic charts, see E. A. Reeves's 1918 article "Halley's Magnetic Variation Charts" in *Geographical Journal*, vol. 51, pp. 237-240.

Even California Governor Arnold Schwarzenegger might enjoy R. V. Tooley's *California as an Island*, from the Map Collectors' Series, no. 8, published in London in 1964.

Henry Coley is quoted in Jonkers's chapter, "Plotting the Third Coordinate," of *Earth's Magnetism in the Age of Sail*. This source also details the assorted tables of magnetic variation available before Halley's voyage.

Flamsteed's 1686 letters to Towneley are available in the Royal Society's manuscript collection but have also been published.

Latin translation by Sydney Chapman's wife, as published in Chapman, 1941.

CHAPTER 10
PRINCE OF TIDES

The journal from Halley's third voyage was never published, or if it was all records of its printing were lost. This makes its interpretation for historians more difficult.

For more on the publication of Flamsteed's *Historia Coelestis*, see Westfall's *Never at Rest* or Christianson's *In the Presence of the Creator*.

A. Chapman, 1982, *The Preface to John Flamstted's "Historia Colestis Britannica,"* National Maritime Museum, Greenwich.

For a concise and complete account of how the clockmaker solved the longitude quandary, see Dava Sobel's rather brilliant *Longitude: The True Story of a Lone Genius Who Solved the Greatest Scientific Problem of His Time.*

The true nature of so-called secular variation wasn't completely understood until the mid-19th century. The ever-so-gradual slowing of Earth's rotation causes the apparent effect. The lunar month appears shorter over centuries because day length increases. Nonetheless, the true effect is similar enough to Halley's description that the astronomical community generally credits him with originally discovering the phenomenon.

The bay of Tonkin's tidal range was proportional to the versed sine of twice the Moon's declination.

"Captains' Letters 1698-1701," Admiralty Archives, Public Records Office, London, quoted by E. F. MacPike, op. cit., p. 116.

More detail on his chart can be found in a 1942 article in *Geographical Journal* by J. Proudman entitled "Halley's Tidal Chart," vol. 100, pp. 174-176.

For more on La Manche, try *English Channel: A Celebration of the Channel's Role in England's History.*

CHAPTER 11
QUEEN ANNE'S PATRONAGE

The translation of Halley's ode to Queen Anne was done by Sydney Chapman's wife and published in his 1941 paper.

For a complete biography on Queen Anne, Green's book on the subject suffices. And for Queen Anne in context, see Cruickshanks's *Glorious Revolution*.

One of the best summaries of Halley's hollow Earth hypothesis is found in a 1992 issue of the *Journal for History and Astronomy*, "The Hollow World of Edmond Halley."

David Kubrin's chapter, "Such an Impertinently Litigious Lady," in *Standing on the Shoulders of Giants* also gives a strong summary of the struggle between Hooke and Halley and Newton. Also see Hooke's *Posthumous Works*.

For more on the impact and distribution of Halley's charts, see W. F. J. Morzer Bruyns, "Longitude in the Context of Navigation," in the *Quest for Longitude*.

A full version of the original Longitude Act is published in Humphrey Quill's *John Harrison: The Man Who Found Longitude*, pp. 225-227. The prize was actually 10,000 pounds for determining longitude within one degree, 15,000 pounds for within 45 minutes, and 20,000 pounds for within 30 seconds.

Ultimately, the 20,000-pound prize was awarded to John Harrison for his marine clock.

After obtaining a certain amount of experience as a commander in the Royal Navy, a captain was "posted" and awarded the title of post-captain.

CHAPTER 12
BACK TO THAT COMET

For an excellent chronicle of the return of Halley's comet in 1985-1986, Peter Lancaster Brown's *Halley's Comet & the Principia* is a recommended read.

In 1729, Halley was elected a foreign member of the Academie de Sciences in Paris.

Some consider Halley's original burial outside a parish church in Lee to be a travesty. The slight was rectified in November 1986 when a memorial at Westminster Abbey was dedicated to Halley.

Norman Thrower first published this translation of the inscription on Halley's capstone.

CHAPTER 13
LEGACY: MORE THAN A COMET MAN

Sir George Clark credited the restoration of Charles II for making the scientific movement fashionable in *The Later Stuarts*.

For a more in-depth view of the history of magnetism, Gerrit Verschuur's *Hidden Attraction: The Mystery and History of Magnetism* is very informative and well done.

For an interesting exposition on the role of Samuel Pepys in building the British Navy, see Arthur Herman's *The World as We Know It Today*.

For further reading on the development of international trade, try P. D. Curtin's *Cross-Cultural Trade in World History* and R. Robert's *Chartered Companies and Their Role in the Development of Overseas Trade*.

W. Doyles's *The Old European Order 1660-1800* was also helpful in writing this epilogue.

In the *Quest for Longitude*, A. J. Turner expresses the importance of failure to scientific progress well in the context of longitude in his chapter, "In the Wake of the Act, But Mainly Before." He writes: "These unsuccessful efforts are just as much part of the development of understanding of the longitude problem as their more successful rivals. And they are just as worthy of study."

Norman Thrower explains Halley's relationship to his patrons eloquently in "The Royal Patrons of Edmond Halley, with Special Reference to His Maps" in *Standing on the Shoulders of Giants*.

The *Boston Gazette and Country Journal* carried the 12-part series on the return of the comet. All 12 parts are published in Craig Waff's "Tales from the First International Halley Watch (1755-59)." Waff's chapter entitled "The First International Halley Watch" in *Standing on the Shoulders of Giants* goes into much greater detail on the comet's return.

S. P. Oliver is quoted from his paper on Halley published in 1880 in *The Observatory*.

ACKNOWLEDGMENTS

I am grateful to many persons over the three years of this project who provided information, insight, and support.

I am indebted to the U.S. Library of Congress for its extensive holdings and for use of a research desk there for several years to complete my work and writings. I wish to express my sincere appreciation to Prosser Gifford, director of the Library of Congress's Office of Scholarly Programs, and the directors of its Rare Book and Special Collections and Science, Technology and Business Divisions and their librarians and staff, in particular, Bruce Martin. The library's vast resources are truly amazing, especially its Chart and Map Collection, where among its holdings I stumbled across Halley's chart of the Atlantic.

I thank the National Maritime Museum, Greenwich, for its gracious assistance with pictures and research, and Rob Warren and Sara Grove in particular.

I am ever thankful for the marvelous holdings at the British Library, especially its newspapers and manuscript collections. It is a vibrant facility with a highly professional staff.

Special thanks to the librarians of the Royal Society and their staff and to Royal Society President Robert McCredie May for allowing me to peruse their collections. Christine Woollett was especially helpful with artwork.

Peter Hingley of the Royal Astronomical Society provided very helpful picture and research assistance.

The Public Records Office, Kew, kept the Admiralty records going back to Halley's day, including the court-martial of Lieutenant Edward Harrison, which is a truly remarkable archiving feat.

I also thank the librarians at the Paris Observatory, particularly Josette Alexandre, and Luisa Pigatto at the INAF-Astronomical Observatory of Padua, Italy, for providing the spectacular image of the fresco of the orbit of Halley's comet.

I owe a debt of gratitude to John Aubrey Adam without whom this book would not have been possible. He faithfully and candidly read chapters, helped me struggle through the tough sections, and pushed me to keep rewriting and rewriting when I thought I was already done.

Special thanks to my wonderful editor, Jeffrey Robbins, at the Joseph Henry Press. His sophisticated sensibilities raised the work another level.

I am grateful to my agent, Patty Moosbrugger, for believing in this project and to Stuart Krichevsky for helping get the book off the ground.

I thank the anonymous reviewers of my work who helped me better capture the nuances of this narrative and better understand English society in Halley's time. I thank the Joseph Henry Press staff, Stephen Mautner, Ann Merchant, and Barbara Kline Pope, especially Sally Stanfield and Jim Gormley, for putting the book together, Michele De La Menardiere for the beautiful cover design, and Estelle Miller and Rachel Marcus. I also thank copyeditor Barbara Bodling O'Hare and photographer Jeremy Whitaker from the Land of Nod for his image of Thornhill's portrait of Sir Isaac Newton.

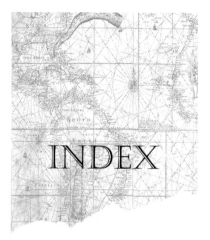

INDEX

Graham, James, 138
Gravitational theory, of Newton, 13
Great Fire of London of 1666, 8, 19, 25,
    72, 76
Greaves, John, 234
Greenwich Observatory, 23, 93, 166,
    185, 193
  founding of, 41, 54, 70, 235
  Halley named Astronomer Royal at,
    163–165
  reequipping, 165
  site of, 7
Gregorian calendar, 229
Gregory, James, 103
Grey, Ralph, 117
Guillotin, Joseph, 138
Gulf of Tonkin, 152–153, 242
Gulliver's Travels, 137, 239
The *Guynie,* 113–114

H

Halley, Captain Edmond, 16. *See also*
    Royal Society of London
  a champion of the Enlightenment,
    187
  charged with improving navigation,
    12, 39, 89–90, 113–114
  close approach to Antarctica, 99–
    100, 104, 191
  contradictions within, 186
  contributions in mathematics, 163
  contributions to geophysics, 142
  critics of, 34
  death of, 178
  difficulties handling his first crew,
    60–63
  diplomacy of, 106, 187–188
  early contacts with Flamsteed, 50,
    55
  elected secretary of the Royal
    Society, 28, 196

first voyage of, 1–79
friend of Isaac Newton, 3, 116, 180,
    183
gaining prominence from
    association with Newton, 31,
    78–79, 171
grieving loss of cabin boy, 84–85,
    87
intelligence collection by, 154, 157
interest in a hollow earth, 174–175,
    243
interest in astronomy, 50, 78
interest in comets, 22, 157 168–174
interest in diving equipment, 114,
    116, 240
interest in geomagnetism, 6, 39, 50,
    133–136, 138, 175
interest in hurricanes, 113
interest in social relationships, 122
interest in the aurora borealis, 175–
    176
journal entries from, 196–197, 203–
    204
laying groundwork for global
    positioning systems, 190
made Astronomer Royal, 163
a meticulous observer, 85, 87–88,
    149–150
most important accomplishments
    of, 104, 188
murder of his father, 71–73
navigational talents of, 12, 209, 231
Pepys complimenting his character,
    12
plan to survey the world magnetic
    field, 6
posthumous honors received by,
    194
predictions regarding comets, 178–
    180
religious stance of, 31–34, 171–172,
    176–177

N

role in building the British Navy, 244
as Secretary of the Admiralty, 89,
187, 236
Peregrinus, Petros, 136–137
Perkins, Peter, 134, 174
Personal hygiene, 120
Peter the Great, 115
Phase transitions, magnetism offering
ways to study, 189
Philosophical Transactions. See Royal
Society of London
Physics, in low dimensions, magnetism
offering ways to study, 189
Pinfold, Richard, captain's servant, 9, 149
"Pinks," 5, 14–15, 38–39, 58, 79
Piracy, 237
danger of, 48
punishment for, 66–68
"Plain Chart" reckoning, 90
Planetary motion, Kepler's laws of, 70–
71, 132
Plato, 26, 31
Plymouth, 63
The *Plymouth,* 149
Polo, Marco, 5
The Pool of London, 11
Pope, Alexander, 26–27
Popery, "abomination" of, 19
Portsmouth harbor, 16–17
Post Man, 20–21
Price, Richard, boatswain, 156
Price, Thomas, carpenter, 9
The *Principia,* 18, 30–32, 76–79, 136,
148, 237–238
second edition of, 171–172
Privateers, 49
Ptolemy, 104–105, 234
Public spirit, British concept of, 231
Published papers, by university
students, 55
Punishment. *See* Discipline on long
ocean voyages
Purcell, Henry, 7

Q

Quadrant, 52, 92
Queen Anne's War, 158, 161
Queen's College, 31, 50
Queiros, Pedro Fernandez de, 5

R

Raleigh, Sir Walter, 185
Random disorder, magnetism offering
ways to study, 189
Ray, John, 142
Reconciliation, between empirical
studies and human values, 191
Religion, dominating English life, 31
Resection method, in surveying, 156
Rio de Janeiro, 91–92, 101, 129
Robinson, Tancred, 109
Robinson Crusoe, 12, 238
The *Roebuck,* 239
Roemer, Olaus, 164
"Round robin" mutinies, 66, 237
Royal African Company, 84, 113
The Royal Mint, 9, 72, 114, 188
Newton's work at, 9
The Royal Navy, 10–11, 59, 65, 69, 89,
118, 231–232
Halley's letters to, 195
Halley's many services to, 188
Royal Observatory. *See* Greenwich
Observatory
Royal Society of London, 11, 28–35,
52, 105–106, 115, 130, 132, 137,
177, 184, 233
criteria for nomination to, 25
establishment of, 230
Halley as clerk of, 34, 70, 188
Halley as secretary of, 140
Halley's election to, 55
Halley's election to the council of,
159